WHY VULNERABILITY STILL MATTERS

We think vulnerability still matters when considering how people are put at risk from hazards and this book shows why in a series of thematic chapters and case studies written by eminent disaster studies scholars that deal with the politics of disaster risk creation: precarity, conflict, and climate change.

The chapters highlight different aspects of vulnerability and disaster risk creation, placing the stress rightly on what causes disasters and explaining the politics of how they are created through a combination of human interference with natural processes, the social production of vulnerability, and the neglect of response capacities. Importantly, too, the book provides a platform for many of those most prominently involved in launching disaster studies as a social discipline to reflect on developments over the past 50 years and to comment on current trends.

The interdisciplinary and historical perspective that this book provides will appeal to scholars and practitioners at both the national and international level seeking to study, develop, and support effective social protection strategies to prevent or mitigate the effects of hazards on vulnerable populations. It will also prove an invaluable reference work for students and all those interested in the future safety of the world we live in.

Greg Bankoff works on community resilience and the way societies adapt to hazard as a frequent life experience. For the last 30 years, he has focused his research on understanding how societies, both past and present, have learnt to normalise risk and the way communities deal with crisis through a historical sociological approach. His publications include co-authoring The Red Cross's *World Disaster Report 2014: Focusing on Culture and Risk*, and a companion, coedited volume entitled *Cultures and Disasters: Understanding Cultural Framings in Disaster Risk Reduction* (2015).

Dorothea Hilhorst focuses on aid-society relations: studying how aid is shaped by the manifold actions of actors in and around programmes for protection, service delivery and capacity development. She has a special interest in the intersections of humanitarianism with development, peacebuilding, and gender-relations. She has done extensive work on humanitarian accountability and situations where disasters meet conflict. Her research programmes have taken place in many settings affected by disaster, conflict, and fragility. Currently, her main research programme concerns changes in humanitarian governance and opportunities for accountability and advocacy, and practices of transactional sex in humanitarian crisis situations.

Routledge Studies in Hazards, Disaster Risk and Climate Change
Series Editor: Ilan Kelman, *Professor of Disasters and Health at the Institute for Risk and Disaster Reduction (IRDR) and the Institute for Global Health (IGH), University College London (UCL)*

This series provides a forum for original and vibrant research. It offers contributions from each of these communities as well as innovative titles that examine the links between hazards, disasters and climate change, to bring these schools of thought closer together. This series promotes interdisciplinary scholarly work that is empirically and theoretically informed, with titles reflecting the wealth of research being undertaken in these diverse and exciting fields.

Disaster Deaths
Trends, Causes and Determinants
Bimal Kanti Paul

Disasters and Life in Anticipation of Slow Calamity
Perspectives from the Colombian Andes
Reidar Staupe-Delgado

The Invention of Disaster
Power and Knowledge in Discourses on Hazard and Vulnerability
JC Gaillard

Earthquakes and Volcanic Activity on Islands
History and Contemporary Perspectives from the Azores
David Chester, Angus Duncan, Rui Coutinho and Nicolau Wallenstein

Why Vulnerability Still Matters
The Politics of Disaster Risk Creation
Greg Bankoff and Dorothea Hilhorst

Health, Wellbeing and Community Recovery in Fukushima
Sudeepa Abeysinghe, Claire Leppold, Alison Lloyd Williams and Akihiko Ozaki

For more information about this series, please visit: https://www.routledge.com/Routledge-Studies-in-Hazards-Disaster-Risk-and-Climate-Change/book-series/HDC

WHY VULNERABILITY STILL MATTERS

The Politics of Disaster Risk Creation

Edited by Greg Bankoff and Dorothea Hilhorst

Routledge
Taylor & Francis Group
LONDON AND NEW YORK

Cover image: © Sino Images / Getty Images

First published 2022
by Routledge
4 Park Square, Milton Park, Abingdon, Oxon OX14 4RN

and by Routledge
605 Third Avenue, New York, NY 10017

Routledge is an imprint of the Taylor & Francis Group, an informa business

British Library Cataloguing-in-Publication Data
A catalogue record for this book is available from the British Library

Library of Congress Cataloging-in-Publication Data
Names: Hilhorst, Dorothea, editor. | Bankoff, Greg, editor. Title: Why vulnerability still matters : the politics of disaster risk creation / edited by Greg Bankoff and Dorothea Hilhorst. Description: Abingdon, Oxon ; New York, NY : Routledge, 2022. | Series: Routledge studies in hazards, disaster risk and climate change | Includes bibliographical references and index. Identifiers: LCCN 2021052655 (print) | LCCN 2021052656 (ebook) | ISBN 9781032113418 (hardback) | ISBN 9781032113432 (paperback) | ISBN 9781003219453 (ebook) Subjects: LCSH: Disaster relief--Political aspects. | Hazard mitigation--Political aspects. | Natural disasters--Political aspects. Classification: LCC HV553 .W49 2022 (print) | LCC HV553 (ebook) | DDC 363.34/8--dc23/eng/20220112
LC record available at https://lccn.loc.gov/2021052655
LC ebook record available at https://lccn.loc.gov/2021052656

ISBN: 978-1-032-11341-8 (hbk)
ISBN: 978-1-032-11343-2 (pbk)
ISBN: 978-1-003-21945-3 (ebk)

DOI: 10.4324/9781003219453

Funded by the European Research Council (ERC), HUMGOV project, grant agreement No 884139.

European Research Council
Established by the European Commission

CONTENTS

List of illustrations *ix*
List of contributors *x*

1 Introduction: why vulnerability still matters 1
Dorothea Hilhorst and Greg Bankoff

PART I
Why vulnerability still matters **13**

2 Remaking the world in our own image: vulnerability, resilience, and adaptation as historical discourses 15
Greg Bankoff

3 Between precarity and the security state: A post-vulnerability view 33
Kenneth Hewitt

4 Creating disaster risk and constructing gendered vulnerability 51
Sarah Bradshaw, Brian Linneker, and Lisa Overton

5 What must be done to rescue the concept of vulnerability? 68
Terry Cannon

PART II
Vulnerability, conflict, and state–society relations **89**

6 Disaster studies and its discontents: the postcolonial state in hazard and risk creation 91
Ayesha Siddiqi

7 Humanitarianism: navigating between resilience and vulnerability 107
Dorothea Hilhorst

8 Resilience, food security, and the abandonment of crisis-affected populations 125
Susanne Jaspars

9 Vulnerability and resilience in a complex and chaotic context: evidence from Mozambique 143
Luís Artur

PART III
Disaster risk creation **169**

10 Power writ small and large: how disaster cannot be understood without reference to pushing, pulling, coercing, and seducing 171
Ben Wisner

11 Disaster risk creation: the new vulnerability 192
Thea Dickinson and Ian Burton

12 Vulnerable Anthropocenes?: towards an integrated approach 206
Kasia Mika and Ilan Kelman

13 'The hottest summer ever!': exploring vulnerability to climate change among grain producers in Eastern Norway 222
Bjørnar Sæther and Karen O'Brien

Index *238*

LIST OF ILLUSTRATIONS

Tables

5.1 Matching DRR actions to reducing the harms from a particular hazard 74
9.1 Vulnerability mapping by score cards (exposure, sensitivity, and adaptive capacity) 154
9.2 Number of years men had spent in South Africa (SA) and their employment 156
9.3 Social protection and vulnerability 158
9.4 Extension services and vulnerability 159
9.5 Seeds and their relevance for local farmers 160
11.1 Vulnerability and disaster paradigms 195
13.1 Weather in the growing season at Årnes 2018. Source: Norwegian Meteorological Institute 227
13.2 Production of wheat and oats in Eastern Norway in tons. Source: Statistics Norway 227

Figures

9.1 Geographical location of Mabote district and its administrative division 153
9.2 Annual rainfall in Mabote 1988–2016 (SDAE 2016) 154
9.3 A plate of locally desired seeds containing millet, sorghum, cowpea, cucumber, and watermelon 160
11.1 Global Insured Disaster Losses from 1970 to 2020 194
11.2 Global Mortality from Disasters 2000 to 2020 194
11.3 Global Mortality from Disasters 2000 to 2020, including losses attributed to SARS-CoV-2 195

CONTRIBUTORS

Luís Artur is a Lecturer and Researcher, Eduardo Mondlane University, Mozambique. Luís works on the sociology of disasters and climate change adaptation. He has been involved with local-led adaptation planning, adaptive social protection, socio-ecology, and M&E systems for disaster risk reduction and climate change adaptation. He works extensively with government institutions, donors, the private sector, and NGOs across several countries, and is former dean of the Faculty of Agronomy and Forestry Engineering at the Eduardo Mondlane University. Amongst others, Artur is a member of the Red Cross movement, of the Peasants Union in Mozambique, and of the International Humanitarian Studies Association.

Greg Bankoff is Affiliate Professor, Ateneo de Manila University and Professor Emeritus, University of Hull. He works on community resilience and the way societies adapt to hazard as a frequent life experience. For the last 30 years, he has focused his research on understanding how societies, both past and present, have learnt to normalise risk and the way communities deal with crisis through a historical sociological approach. His publications include co-authoring *The Red Cross's World Disaster Report 2014: Focusing on Culture and Risk* and a companion, coedited volume entitled *Cultures and Disasters: Understanding Cultural Framings in Disaster Risk Reduction* (2015).

Sarah Bradshaw is a Professor of Gender and Sustainable Development and Head of the School Law, Middlesex University. A feminist and scholar-practitioner, Sarah combines academic work with advocacy activities. She wrote the background paper on gender for the High-Level Panel that developed the first draft of the SDGs. She has published widely on the theme of gender and development, and gender and disasters, in 2013 she published the first book to focus on the nexus of *Gender, Development and Disasters* (Edward Elgar). She is thematic lead on 'Intersectionality and Gender' for the 'Gender Responsive Resilience and Intersectionality in Policy and Practice' (GRRIPP) project and is currently working on a second book for Edward Elgar *Rethinking Gender and Disasters*.

Ian Burton is a Professor Emeritus at the University of Toronto. Independent scholar and consultant, specialist in human adjustment to natural hazards and adaptation to climate change and extreme events. Working on risk assessment and management, and science for public policy. Director, Institute for Environmental Studies, University of Toronto, 1979-84; Senior Policy Advisor and Director Adaptation Research Group, Environment Canada, 1988-1992; Lead Author for three IPCC Assessments; consultant assignments with Ford Foundation, UNESCO, World Bank, World Health Organization, and numerous national and international agencies and consulting companies. Fellow, Royal Society of Canada. Member World Academy of Arts and Science; Munich Re Climate Insurance Initiative; Institute for Catastrophic Loss Reduction, Canada. See http://www.burtoni.ca

Terry Cannon is a Senior Research Fellow at Institute of Development Studies (IDS) at University of Sussex. Terry teaches postgrad courses on climate change at IDS and on disaster vulnerability and climate change adaptation at King's College London and several other European universities. A development studies specialist, his current research is on culture and risk, the social construction of vulnerability, and disaster risk creation. He was co-editor and author of the IFRC World Disasters Report 2014: focus on culture and risk and the related book Cultures and Disasters (Routledge 2015). He has been affiliated with the International Centre for Climate Change and Development (ICCCAD) in Dhaka for many years and worked on disasters and climate change in Bangladesh.

Thea Dickinson is an Independent Consultant. She is a climate change adaptation specialist. In 2019, she completed her PhD in Physical and Environmental Science at the University of Toronto with an emphasis on identifying determinants of national-level climate change adaptation. She was contributing author for the IPCC Working Group II and the Special Report on Extremes. Her current research interests include successful adaptation strategies, the disaster-climate policy interface, and disaster risk creation. Since 2007 has authored and co-authored over 30 publications on the human dimensions and societal responses to climate change and disasters.

Kenneth Hewitt is Professor Emeritus, Department of Geography and Environmental Studies & Research Associate, Cold Regions Research Centre, Wilfrid Laurier University, Waterloo, Ontario. Kenneth's academic career includes a PhD in Geomorphology from Kings College, London University, and tenured professorships at the University of Toronto, and Rutgers University, New Jersey. His main teaching, research, and consulting fields are in high mountain geomorphology and environments, critical disaster theory and practice, and peace research. Landform interests are in glacial geomorphology and hydrology, and special attention to surge-type glaciers, catastrophic landslides, natural dams, and outburst floods. He has looked at disaster concepts, vulnerability, and adaptation to environmental crises. His peace research mainly concerns impacts of armed violence on civilians, cities, and habitats. He has regional specialisations in mountain environments mainly the Karakoram Himalaya, Inner Asia, having spent 17 field seasons there. He has published extensively in these fields.

Dorothea Hilhorst is Professor of humanitarian studies at the International Institute for Social Studies (ISS) of Erasmus University in The Hague. Dorothea Hilhorst focuses on aid-society relations: studying how aid is shaped by the manifold actions of actors in and around programmes for protection, service delivery, and capacity development. She has a special interest in the intersections of humanitarianism with development, peacebuilding, and gender-relations. She has done extensive work on humanitarian accountability and situations where disasters meet conflict. Her research programmes have taken place in many settings affected by disaster, conflict, and fragility. Currently, her main research programme concerns changes in humanitarian governance and opportunities for accountability and advocacy, and practices of transactional sex in humanitarian crisis situations.

Susanne Jaspars is an Independent Researcher and Research Associate at the Food Studies Centre of the School for Oriental and African Studies. Susanne Jaspars has researched the social and political dynamics of food security in situations of conflict, famine, and humanitarian crises, for more than 35 years. She has worked mostly in the Horn of Africa, often Sudan. In 2018, she published her PhD as a book entitled *Food Aid in Sudan. A History of Politics, Power and Profit* (Zed Books), adding to a range of other books, academic articles, and policy reports.

Ilan Kelman is Professor of Disasters and Health, University College London, UK and Professor II, University of Agder, Norway. Ilan is on Twitter/Instagram at @ILANKELMAN and has overall research interests of linking disasters and health, including the integration of climate change into disaster research and health research. This work covers three main areas: (i) disaster diplomacy and health diplomacy http://www.disasterdiplomacy.org; (ii) island sustainability involving safe and healthy communities in isolated locations https://www.islandvulnerability.org/; and (iii) risk education for health and disasters https://www.islandvulnerability.org/.

Brian Linneker is an Independent Scholar and Senior Research Fellow in Economic Geography and the Environment. Brain has worked for 30 years on sustainable urban and rural development, and poverty, vulnerability, and social exclusion for UK government departments, UK international and Latin American national NGOs and civil society organisations, and within various academic institutions including the London School of Economics, King's College London, Birkbeck College, Queen Mary University and Middlesex University. He has published over 100 articles, reports, book chapters, and working papers on London, the UK, and Latin America.

Kasia Mika is Lecturer in Comparative Literature at Queen Mary University London. Kasia previously worked at the University of Amsterdam and the

Royal Netherlands Institute of Southeast Asian and Caribbean Studies (KITLV). Her monograph, *Disasters, Vulnerability, and Narratives: Writing Haiti's Futures* (Routledge 2019), turns to concepts of hinged chronologies, slow healing, and remnant dwelling, offering a vision of open-ended Caribbean futures. She produced *Intranqu'îllités* (dir. Ed Owles, Postcode Films; AHRC Research in Film Award) and co-authored *(Un)timely Crises: Chronotopes and Critique* (Palgrave 2021). She published in *The Journal of Haitian Studies*; *Moving Worlds*; *Modern and Contemporary France*; *Area Journal*.

Karen O'Brien is Professor in Department of Sociology and Human Geography at the University of Oslo, Norway. Karen O'Brien is co-founder of cCHANGE, an organisation that translates research into action for transformative sustainability results. Karen's research emphasises the social and human dimensions of climate change, including the relationship between climate change adaptation and transformations to sustainability. She is interested in the role of beliefs, values, worldviews, and paradigms in generating conscious social change. Her recent books include *You Matter More Than You Think: Quantum Social Change for a Thriving World* and *Climate and Society: Transforming the Future* (co-authored with Robin Leichenko).

Lisa Overton is Senior Lecturer in Politics at Middlesex University. Lisa works on gender and disasters focusing particularly on the intersections with youth and sexualities through oral and life history storytelling. She has worked on issues related to women's access to housing and land in post-tsunami Sri Lanka, young women and gender diverse young peoples' reflections on identity in Post-Katrina New Orleans, notions of collectivity and change during the UCU-USS strikes in 2018 and most recently is examining the crisis of consent, complicity and the meaning of sexual misconduct in higher education and queer spaces.

Bjørnar Sæther is Professor in Human Geography, Department of Sociology and Human Geography at the University of Oslo. Bjørnar Sæther has specialised in economic geography and has written extensively on rural development, innovation in food, agriculture, and other natural resource-based industries. Questions concerning nature, environment, and climate change have always been part of his research and teaching interests.

Ayesha Siddiqi is Assistant Professor in the Department of Geography at the University of Cambridge. She is a postcolonial and development geographer who has worked for over a decade in municipalities and neighbourhoods affected by hazard-based disasters, dealing with underlying issues of conflict and insecurity in the Global South. Her book *In the Wake of Disaster: Islamists, the State and a Social Contract in Pakistan* was published by Cambridge University Press in 2019 and she has worked in policy with the UK's Houses of Parliament, UNDRR, and various think tanks.

Ben Wisner is an Honorary Visiting Professor, Institute for Risk and Disaster Reduction, University College London, UK & Affiliate Scholar, Environmental Studies Program, Oberlin College, Oberlin, OH, USA. Ben has worked for

56 years on the use of village and neighbourhood vernacular knowledge in resource and risk management. He also studies whether and how national and international policy and practice acknowledge, respect, and learn from villages and neighbourhoods. His works include *Power and Need in Africa* (Earthscan/ Africa World Press, 1988/ 1989); *At Risk: Natural Hazards, People's Vulnerability and Disasters* (co-authored (Routledge, two editions, 1994 & 2004); *The Routledge Handbook of Natural Hazards and Disaster Reduction* (edited with two others, 2012); *Vulnerability as Concept, Model, Metric and Tool* (Oxford University Press online, 2016 https://oxfordre.com/naturalhazardscience/view/10.1093/acrefore/9780199389407.001.0001/acrefore-9780199389407-e-25).

1

INTRODUCTION

Why vulnerability still matters[1]

Dorothea Hilhorst and Greg Bankoff

We think vulnerability still matters or, at least, we think it matters to ask the question of whether it still matters. Vulnerability has been the key concept of disaster studies for a long time. What may be dubbed the 'iron law' of disaster studies stipulates that disasters cannot be equated to the hazard (Wisner et al. 2012), but are the outcome of hazards encountering vulnerability, mitigated by response capacities (Wisner et al. 2004). Whether a disaster unfolds as a consequence of an earthquake, for example, depends on poverty levels in the population and the state of the built environment (Kelman 2020; Wisner et al. 2012). The power of the concept of vulnerability has been that it explains the differentiated impact of hazards and highlights the socially constructed nature of vulnerability – and hence of disasters – as produced by politics, economic processes, and social exclusion (Bankoff et al. 2004).

However, since the turn of the millennium, the central status of vulnerability has been largely overtaken by another concept, namely resilience. The UNDRR defines resilience as the ability of a system, community, or society exposed to hazards to resist, absorb, accommodate, adapt to, transform, and recover from the effects of a hazard in a timely and efficient manner. Resilience has been welcomed because it focuses on people's agency and the capabilities of communities to withstand disaster rather than the supposed passivity associated with the term vulnerability. It was seen as the flipside of vulnerability: the less vulnerable the more resilient and vice versa. Nonetheless, the authors in this volume think the term vulnerability matters. We believe that case analyses informed by vulnerability lead to radically different political outcomes than approaching the same cases from a lens of resilience.

Nearly 20 years ago, a group of concerned scholars, motivated by a shared commitment to improving people's ability to resist disasters, came together in a workshop to discuss the utility of vulnerability as both a conceptual and

DOI: 10.4324/9781003219453-1

methodological tool. That workshop led to the publication of *Mapping Vulnerability: Disaster, Development and People* in 2004. In it, the authors stress the social production of vulnerability, or how the relative position of advantage or disadvantage that a particular group holds within a society's social order, renders it unsafe. Many things have changed since then and particularly since vulnerability first became popularised as a term in the 1970s – both in the world and about how we think about the world.

For a start, it is a very different world now, politically, with the end of the Cold War and the demise of the Soviet Union. The liberal institutions upon which the post-World War II international system was built, the so-called Bretton Woods agreement, is bursting at the seams to accommodate the rise of new nationalisms. Inequality within and between nations has increased as the gap between rich and poor has widened in the last 30 years (Freeland 2012). The workplace, too, is not the same with the casualisation of employment and the rise of zero-hours contracts. Actually, of course, these conditions have long been the norm in lower- and middle-income countries (LMICs), and the novelty is in their application to the hitherto largely protected labouring classes of the industrialised world. Over the last 20 years, disasters have become more frequent and wrought greater impact. Disasters claimed approximately 1.23 million lives, affected over 4 billion people, and led to about US$ 2.97 trillion in economic losses worldwide between 2000 and 2019 (UNDRR 2020). Climate change has become the central focus of our concern. Indeed, the present age has even been renamed the Anthropocene (Crutzen and Stoermer 2000).

At the same time, people have begun to be considered not just as vulnerable, a condition seen as emphasising their frailties, but also as resilient, with the capacities to organise, resist, learn, change, and adapt. In fact, the latter, adaptation, has become the slogan and the solution of just how human systems need to adjust to actual or expected climatic stimuli. While vulnerability has certainly not disappeared from the global political and conceptual arena, its influence over the way that disasters are thought about and managed has waned considerably in recent years.

We think it's time to re-centre the debate on vulnerability and the processes that continue to make or are even creating new risks for peoples and societies. Resilience and adaptation are welcome additions in understanding disaster risk reduction, but they also serve (and have been used as such) to distract attention from certain key issues. Resilience can be used to mask social inequalities and absolve states from their duty of care, and adaptation runs the risk of blaming disasters on nature once again and returning to an older and all too familiar hazard-focused trope. Vulnerability still places the stress where it rightly should be – on the processes (political, social, economic, and environmental) that put people at risk (Cannon, this volume). It is time to restate this and, at the same time, perhaps, expand on our earlier conceptualisations. As the world changes, do we need to rethink what vulnerability means and how we apply it? Hence, a second workshop and a second volume, that includes some of the original contributors, to see how their ideas have changed, as well as to give space to fresh voices that may see vulnerability through new eyes and from alternate perspectives.

Why vulnerability still matters

Vulnerability, when it was first proposed, was a novel way of understanding why disasters happen and who was hurt by them. It was also a critique of the political and socio-economic systems that produced them. Its proponents implicitly blamed disasters on the legacy of colonialism and the neo-colonial forms of domination exercised by western governments and transnational corporations over the newly independent nation-states of the now denominated Third World (O'Keefe, Westgate, and Wisner 1976; Watts 1983). Without overtly taking sides in the Cold War debates between capitalism and communism, though some scholars came close to an ideological position (Cannon 1994), the proponents of vulnerability confined their criticisms to the practices of transnational capital and the unfavourable terms of trade they imposed upon poorer nations. They argued that communities were rendered unsafe by the relative position of advantage or disadvantage that particular groups occupied within a society's social order (Hewitt 1997, 141). Vulnerable people were at risk, not simply because they were exposed to hazard, but as a direct result of a combination of variables such as class, caste, ethnicity, age, gender, and disability (Wisner 1993, 131–133) that determined people's entitlements and affected their command over basic necessities and rights (Hewitt 1997, 143–151; Watts 1993, 118–120). Vulnerability became the defining conceptual framework through which disasters were understood, an approach embodied in the publication of *At Risk: Natural Hazards, People's Vulnerability and Disasters* in 1994, and gained widespread acceptance with the UN's adoption of Resolution 44/236 (22 December 1989) declaring An International Decade for Natural Disaster Reduction beginning on 1 January 1990.

Today, however, we live in a different world. Just how different a world is one of the main issues addressed by the authors in this book. Much has changed since the 1970s. Politically, the Cold War has ended – or was at least suspended – with the disintegration of the Soviet Union in 1991 and the entire international system, the Three Worlds model, upon which it was based (Bankoff, this volume). This categorisation cast countries as either capitalist, industrial, and sharing similar plural political institutions (First World), or as one-party states with industrial or rapidly industrialising centrally planned economic systems (Second World). All other countries were designated as Third World, a highly diverse mix of states culturally and economically whose unifying factor was their complete lack of international influence (Harris, More, and Schmitz 2009, 11).

This tripartite division of the world was coined by Alfred Sauvy, a French demographer and economic historian, who identified the Third World with underdevelopment (Sauvy 1952). If this categorisation was initially based on political distinctions, over time, it increasingly came to signify economic differentiation too. Moreover, these same regions were also those most associated with the occurrence of what were labelled at the time as 'natural disasters' (Bankoff 2001). As the Second World faded into oblivion post-1991, the residual categorisation simply denoted the difference between developed and developing nations: rich and poor, North and South, HICs (high-income countries) and LMICs. Now, it is argued that these 'old labels' no longer serve a useful purpose in summarising either the structural characteristics of national economies or the patterns of interaction between contemporary states. Even the distinction between HICs and LMICs has become increasingly tenuous given the expansion of the number of countries that do not fit into either designation, some of which, such as Russia and China, are regarded as 'poor and powerful, having middle to low incomes but fast-growing economies and considerable geopolitical influence' (Harris, More, and Schmitz 2009, 14).

Socially, too, the world is a very different place than it was in the 1970s. A new class-in-the-making is emerging out of the neoliberalism that underpins the globalisation of the last 30 years – the precariat. It consists, according to Guy Standing (2011), of 'a multitude of insecure people, living bits-and-pieces lives, in and out of short-term jobs, without a narrative of occupational development'. It includes under- or unemployed educated youth, growing numbers of the criminalised, those categorised as disabled, and migrants in their hundreds of millions around the world. It also includes women abused in oppressive labour (Bradshaw et al this volume).

This class has long been in the making in LMICs where the industrial revolution proved to be not so much a stage in the modernisation narrative as a myth that came and went in a single lifetime (Ferguson 1999). The decline in living standards and the rise in poverty levels which accompanied the neoliberal structural adjustment policies of the 1980s/90s in many parts of Sub-Saharan Africa, Latin America, and the former Soviet Union represented a humiliating process of expulsion from the development process, an 'abjection' where the term implied not simply being cast aside but being thrown down as well (Ferguson 2007). What is new, however, is the growth of the precariat in HICs, particularly following the banking crisis of 2007 (Hewitt, this volume). The shock to the financial system gave employers an opportunity to dismiss permanent salaried staff and replace them with new labour arrangements. It also allowed the state to 'reform' welfare benefits. Subsequently, there has been a rise in the number of temporary and agency labour, outsourcing, occupational dismantling, and the abandonment of non-wage benefits by firms, while governments have acted swiftly to erode state benefits and pensions (Standing 2014, 43–99).

A significant proportion of the precariat is the number of forcibly displaced people, currently estimated at 79.5 million worldwide. These are individuals forced from their homes due to persecution, conflict, and human rights violations. Most are IDPs (Internally Displaced Persons), people forcibly uprooted but who remain inside their own country. An increasingly large proportion, however, are refugees (29.6 million in 2020), displaced persons who have crossed international frontiers and who are currently residing in a state in which they cannot claim citizenship (UNHCR 2020). These are persons fleeing conflict and/or persecution in their own country. Added to these is the much larger number of economic migrants, people fleeing destitution and environmental degradation, estimated now at more than 272 million or 3.5 per cent of the world's population (IOM 2020, 2). By abjuring their citizenship and abandoning their state, these transboundary migrants lose their political status. As 'non-citizens', they inhabit an indeterminate zone of disorder and chaos where the normal rules of play do not apply. Currently, only one-third of migrants leave poor countries for rich ones, but migration patterns seem set to change in the decades ahead, driving more people to seek shelter in temperate zones where the effects of climate anomalies are considered less severe (White 2011). Like the precariat of which they are part, migrants are very vulnerable to the impact of environmental change and natural hazards.

But, perhaps, the biggest change in the last 50 years has been environmental. Climate change, only a distant rumble in the 1970s, is now a real and pressing concern. Overwhelming scientific evidence supports the conclusion that human activity is changing the climate and will continue to affect it for hundreds if not thousands of years to come. World temperatures have fluctuated in the past, but it is estimated that at no time in the last 800,000 years have concentrations of greenhouse gases (carbon dioxide, methane, and nitrous oxide) been so high as they are today (Giddens 2014, 12; World Bank 2015, 16). While a significant minority of sceptics remain in some countries (Buchholtz 2020), the current debate is more about whether climate change is gradual, allowing societies sufficient time to adjust (six IPCC reports), or whether it is non-linear, whereby crossing some threshold will precipitate sudden and catastrophic change (Lovelock 2006). There is little that can be done in the latter case apart from ensuring a reduction in greenhouse gases to prevent the tipping point from being reached. If, alternatively, climate change is relatively slow, then individuals and societies will have time to adapt given the necessary inducements and incentives (Szerszynski and Urry 2010, 1–2).

The key term in this discourse is adaptation (IPCC 2007). However, this preoccupation with how well societies will adapt to the increasing intensity of storms, rising sea levels, and more frequent floods runs the risk of blaming Nature once more for disasters and returning to an older hazard-focused paradigm that ignores how such events are socially as well as physically constructed. This is an all too familiar trope that seeks to render large parts of the world as vulnerable by blaming the poverty of these regions squarely on natural forces and disasters on people's lack of resilience (Bankoff 2001). It also serves to diffuse the opprobrium that might otherwise be directed at an economic system created by and, until recently, mainly benefiting western industrialised nations by making declarations that stress the common plight of humanity and by focusing on scientific and technical discussions about purely climatic and scientific phenomena (Bankoff and Borrinaga 2016).

There is no reason to suppose, therefore, that societies today are any less vulnerable than they were in the 1970s. In fact, the evidence suggests quite the opposite: that people are even more at risk than they were before. Inequalities are deepening rather than narrowing, both within and between nations. For more than 71 per cent of the world's population, income inequality is growing and, though inequality among some countries (notably China) is declining in relative terms, the differences among countries are still considerable. The average income of people living in North America is 16 times higher than that of people in sub-Saharan Africa (UN 2020, 3). The nature of this inequality is not only income- and asset-related but is also manifest in the areas of automation, digital innovation, and artificial intelligence. Put simply, there are more losers than winners from developments over the last 50 years. The number of people in the precariat or who feel they could fall into it at any time has swelled to alarming proportions, and the upward trend is likely to continue (Standing 2014, 43–99). The political, economic, and social precariousness of people in this emergent class leaves them less resilient and more vulnerable to the impact of natural hazards and climate change.

Underlying the growing inequality and the nature of precarity is the industrialised West's reluctance to fully acknowledge responsibility for causing climate change. The governments and peoples of HICs are unwilling to accept that their overdevelopment has been at the expense of the underdevelopment of the rest of the world whose peoples are now expected to pay an inordinate share of the socio-environmental consequences of the resultant climate variation. Rather than being a unifying issue around which humanity might rally in the face of a common challenge, climate change is as divisive as any Cold War discourse – though the geographical fracture lines are now more likely to be depicted as North-South rather than East-West. Moreover, the talk is no longer about who is vulnerable so much as who is resilient, what is sustainable and how to adapt. The word vulnerability, for instance, does not appear even once among the 17 Sustainable Development Goals adopted by the United Nations in 2015.

Changing views on vulnerability and disaster risk creation

The original conception of vulnerability invited questions of why people were vulnerable. The famous pressure and release model, for example, unravelled how unsafe conditions (such as living in a dangerous place) were shaped by dynamic pressures (such as urbanisation and the loss of biodiversity) that in turn were related to root causes (such as exclusive institutions or lack of access to democratic representation) (Blaikie et al. 1994). In subsequent decades, vulnerability has increasingly been conceptualised as a condition in which people find themselves, in a non-relational and non-causational static kind of way (Heijmans 2012). According to this approach, vulnerability does not lead to probing questions about the origins of risk but is expressed in numbers and statistics – and reduced by interventions. A striking example of this is described by Susanne Jaspars (this volume) where the vulnerability of people in Sudan is no longer understood against the backdrop of war but simply explained away in terms of personal knowledge gaps and cultural habits. Understanding vulnerability as a personalised and local expression leads to disaster risk reduction interventions that are equally restricted in scope. In Afghanistan, for instance, occasional small walls are built to protect some villages from river flooding. These small-scale measures may be effective to do as designed yet their significance fades in comparison to the larger vulnerabilities faced by communities or in relation to the national situation (Mena and Hilhorst 2020).

As the term vulnerability was gradually denuded of its critical potential, the gap was partly filled by the introduction of another term: disaster risk creation (DRC). The introduction of the concept of disaster risk creation as a corollary to disaster risk reduction may prove to be an effective way of refocusing attention on the root causes of disaster risks (Lewis and Kelman 2012). Disaster risk creation re-emphasises the causes of disaster risks, and, importantly, recognises that disaster risks are created through human interference with natural hazards, the social production of vulnerability, the neglect of response capacities, or a combination of all three. Voters that favour lower taxes in hurricane-prone Texas, for example, can be seen as voting to reduce disaster preparedness – and therefore indirectly choose for disaster risk creation (Kelman 2020). Chapters in this book highlight different aspects of disaster risk creation and vulnerability. Thea Dickinson and Ian Burton outline how a compartmentalised notion of disaster masks how different types of disasters, including natural hazards and pandemics, co-occur and are triggered by similar causes commonplace to our era of the Anthropocene. Bjørnar Sæther and Karen O"Brien argue that many of the previous distinctions between North and South have given away to more globalised and uneven patterns of vulnerability, a point they illustrate in a discussion of the hottest summer ever in Eastern Norway. The chapter by Kasia Mika and Ilan Kelman further explores this issue and outlines how vulnerability must be seen in its temporal dimension, as uneven and playing out at multi-scalar levels.

Early proponents of the vulnerability thesis have been criticised for the unidirectional way in which they viewed local vulnerability as the outcome of global forces (Saddiqi, this volume). Global factors indeed impact local vulnerabilities, but these are also locally produced and, importantly, it is at the local level where global factors are embedded, altered, and instrumentalised in local-level inequalities. The chapter of Ben Wisner builds around a series of case studies to elaborate that power needs to be writ large and small – comprising global forces and local power differentials – to fully comprehend disaster risk creation.

The idea of disaster risk creation can also shed light on how policies that are intended to reduce disaster risk have the opposite effect and enhance or (re)-create disaster risks. Disaster risk reduction activities may lead to social conflict over questions such as who is protected and who pays the price. They may also miss out on their objective to assist the most vulnerable as the benefits are reaped by more affluent and powerful people. An example drawn from this volume concerns the social protection programmes in Mozambique that, due to a technocratic approach, protect the well-to-do at the expense of the most vulnerable

(Artur this volume). Indeed, history is replete with development projects and disaster risk reduction initiatives that lead to further disaster risks, with river dams and mining areas among the notorious examples of what Philippine activists call 'development aggression' (Heijmans 2004).

Vulnerability, conflict, and state–society relations

One of the most remarkable omissions of the early generation of disaster studies is the lack of attention to the impact of war, conflict, and violence. In the realm of high politics, where the Hyogo and Sendai Frameworks for action have been negotiated, references to conflict are entirely absent. This is despite the glaring realities of how conflict affects the frequency and intensity of disasters. There is a strong statistical relation between the occurrence of conflict and disasters, and there have been years when almost every single country experiencing conflict also experienced a severe disaster (Caso, 2019). There is also a strong intuitive and deductive relationship between conflict and disaster, as there are many ways in which conflict can negatively affect the occurrence of hazards such as mismanaging land and water, increasing vulnerabilities through displacement, damaging infrastructure and loss of lives and livelihoods, and diminishing the capacities of people to respond to disaster or reduce disaster risks (Hilhorst, this volume).

The devastations of war and how they have produced vulnerability in World War II is elaborated by Ken Hewitt (this volume). Conflict, however, has many more manifestations that are, perhaps, less visibly devastating yet may be equally impactful when it comes to the production of vulnerability. Even the threat of violence, that is a daily reality in many conflict-affected or authoritarian settings, can be seen to increase vulnerability as people have no legal or other recourse when their resource base is undermined or taken away from them. Vulnerability is also affected by structural and cultural violence. Structural violence refers to processes whereby one groups oppresses another through structural means, such as an exclusion from educational or economic opportunities or restrictions of freedom of assembly and speech by policy or law. Cultural violence is the broader, semi-permanent state through which some forms of physical violence and structural violence are considered as legitimate (Galtung 1996).

Gender and race are among the most prominent institutions that exemplify how cultural violence can impact vulnerability (Bradshaw et al, this volume). These different types of violence often 'co-operate' or work in tandem. In many instances, violence does more to characterise state–society relations than the benevolent idea of a social contract, implicit in the international frameworks for disaster response, in which the state is supposed to protect its citizens. And this brings us back full circle to the importance of considering disaster risk creation. For many people, the state represents a disaster risk multiplier, that is far more impactful than the occasional disaster risk reduction project that comes their way. Taking a comprehensive view on violence and conflict reveals how large parts of the globe, that have been subject to colonial and post-colonial developments, continue to be inherently violent, full of conditions of displacement, and replete with disaster risks (Siddiqi, this volume).

Doing vulnerability and resilience

The question remains how the twin concepts of vulnerability and resilience are being used in the everyday politics of humanitarian action. It is precisely in those conflict-ridden areas where humanitarian operations are realised that it is possible to see how policies that promote resilience and vulnerability are implemented in practice – in actual projects on the ground. Where aid is organised as resource transfers to individuals and families rather than through projects directed at the community level, vulnerability remains central to humanitarian practice and continues to be a prime indicator in determining who will or will not receive such assistance. Susanne Jaspars reviews how in past decades resilience has replaced previous needs-assessment as the organising concept in the design of food security practices during protracted crises. These resilience approaches start from the premise that communities owe their vulnerability largely to their own unhygienic habits, and interventions consequently focus on small-scale improvements in customs or the introduction of food supplements, couched in a language of adapting to climate change and linking relief to development. What is missing is the reality check of whether people and communities are indeed resilient and whether interventions indeed result in improved food security. The reality is somewhat different as the conditions of conflict erode any possibility for food security. What is more, the perverse outcome of these interventions means that communities are often abandoned by the aid community and left to their fate. There is nobody to notice their plight as the people responsible are blinded by the myth of resilience.

Dorothea Hilhorst argues for the importance to research the everyday ways in which humanitarian actors deal with the governance of resilience and vulnerability. First, she elaborates the importance of meso-level analysis of how governance arrangements that evolve around disasters in different political settings deal with resilience and vulnerability. Second, she calls for more studies on the technologies of humanitarian action. Despite some recognition of the importance of the nitty-gritty workings of indicators, categories, and algorithms, critical research is scant. As a result, technological applications can run away with themselves in the absence of the necessary checks and balances inherent in value-driven accountability.

***

In *Mapping Vulnerability*, we asked what makes people vulnerable and then proceeded to describe that landscape of vulnerability, showing that, while there were possible paths on this map, that there were no set routes or even fixed destinations. Two decades later, we find ourselves asking whether vulnerability still matters when, if anything, the terrain is even more complex now than it was then, and we are even more in need of a compass to guide us in the right direction – to ask the hard questions. What's more, it seems that we have decided to abandon a trusted set of coordinates for newer ones; resilience and adaptation create a somewhat different topography of risk. We maintain, however, that vulnerability still matters very much. The poor, the destitute, the disadvantaged, the displaced, the disabled, the un- or underemployed, and the homeless have not disappeared. If anything, the root causes of such conditions are more acute today than they were 50 or even 20 years ago.

The nature of vulnerability, too, has changed/is changing, as the authors in this volume make clear. There is a new political order that is more threatening than the structured certainties of the Cold War past, in which environment-fuelled conflicts or post-conflict situations are considered the norm. There is a new social order, also premised on uncertainties, in which precarity in life opportunities, both personal and professional, is becoming the dominant condition. And, then, there is climate change, the biggest uncertainty of them all, where fluctuations in average temperatures and rainfall mean more floods, more droughts, higher seas, worse storms and much more suffering. While the emphasis over the last 20 years has been on disaster risk reduction, the outcome has often been disaster risk creation. Yes, we need to be resilient. And, yes, we need to adapt. But more than anything else, we need to recognise what makes us vulnerable and address the issues that put us all at risk – some more so than others.

Notes

1 The workshop leading to this volume was supported by the Netherlands Organisation of Scientific Research (NWO),grant number 453-14-013.

References

Bankoff, Greg. 2001. "Rendering the World Unsafe: 'Vulnerability' as Western Discourse." *Disasters* 25 no. 1: 19–35.

Bankoff, Greg, Georg Frerks and Dorothea Hilhorst, eds. 2004. *Mapping Vulnerability: Disaster, Development and People*. London: Earthscan.

Bankoff, Greg, and George Emmanuel Borrinaga. 2016. "Whethering the Storm: The Twin Natures of Typhoon Haiyan and Yolanda." In *Contextualizing Disasters*, edited by Gregory Button and Mark Schuller, 44–65. New York: Berghahn.

Blaikie, Piers, Terry Cannon, Ian Davis, and Ben Wisner. 1994. *At Risk. Natural Hazards, People's Vulnerability and Disasters*. London and New York: Routledge.

Buchholtz, Katharina. 2020. "Climate Change: Where Climate Change Deniers Live." *Statista*. https://www.statista.com/chart/19449/countries-with-biggest-share-of-climate-change-deniers/.

Cannon, Terry. 1994. "Vulnerability Analysis and the Explanation of 'Natural Disasters'". In *Disasters, Development and Environment*, edited by Ann Varley, 13–30. Chichester: John Wiley & Sons.

Caso, Nicolas. 2019. Human development, climate change, disasters and conflict: Linkages and empirical evidence from the last three decades. MA thesis. The Hague: International Institute of Social Studies of Erasmus University Rotterdam. Retrieved January 12, 2022 from: Erasmus University Thesis Repository: Human development, climate change, disasters and conflict: linkages and empirical evidence from the last three decades (eur.nl).

Crutzen, Paul J., and Eugene F. Stoermer. 2000. "The 'Anthropocene'." *Global Change Newsletter* 41 (May): 17–18.

Ferguson, James. 1999. *Expectations of Modernity: Myths and Meanings of Urban Life on the Zambian Copperbelt.* London: University of California Press.

Ferguson, James. 2007. *Global Shadows: Africa in the Neoliberal World Order.* London: Duke University Press.

Freeland, Chrystia. 2012. *Plutocrats: The Rise of the New Global Super-Rich and the Fall of Everyone Else.* New York: Penguin Press.

Galtung, Johan. 1996. "Part II: Conflict Theory." In *Peace by Peaceful Means: Peace and Conflict, Development and Civilization*, edited by Johan Galtung, 70–80. Oslo: London; Thousand Oaks, CA: International Peace Research Institute; Sage Publications.

Giddens, Anthony. 2014. *The Politics of Climate Change.* Cambridge: Polity Press.

Harris, Diane, Mick More, and Hubert Schmitz. 2009. *Country Classification for a Changing World, IDS Working Paper 326.* Brighton: Institute of Development Studies.

Heijmans, Annelies. 2004. "From Vulnerability to Empowerment." In *Mapping Vulnerability: Disasters, Development and People*, edited by Greg Bankoff, Georg Frerks, and Dorothea Hilhorst, 115–127. London and Sterling, VA: Earthscan.

Heijmans, Annelies. 2012. Risky Encounters. Institutions and interventions in response to recurrent disasters and conflict. PhD dissertation. Wageningen University. https://www.wur.nl/en/Publication-details.htm?publicationId=publication-way-343233383734.

Hewitt, Kenneth. 1997. *Regions of Revolt: A Geographical Introduction to Disasters.* Edinburgh: Longmans.

IOM. 2020. World Migration Report 2020. Geneva: International Organisations of Migration. https://www.un.org/sites/un2.un.org/files/wmr_2020.pdf.

IPCC. 2007. *Climate Change 2007 Synthesis Report.* Geneva: Intergovernmental Panel on Climate Change.

Kelman, Ilan. 2020. *Disaster by Choice: How Our Actions Turn Natural Hazards into Catastrophes.* Oxford: Oxford University Press.

Lewis, James, and Ilan Kelman. 2012. "'The Good, The Bad and The Ugly: Disaster Risk Reduction (DRR) Versus Disaster Risk Creation (DRC)'." *PLoS Currents Disasters*, 21 June. doi:10.1371/4f8d4eaec6af8.

Lovelock, James. 2006. *The Revenge of Gaia: Why the Earth Is Fighting Back – and How We Can Still Save Humanity.* London: Allen Lane.

Mena, Rodrigo, and Dorothea Hilhorst. 2020. "'The (Im)possibilities of Disaster Risk Reduction in the Context of High-Intensity Conflict: the Case of Afghanistan.'" *Environmental Hazards*, doi:10.1080/17477891.2020.1771250

O'Keefe, Phil, Ken Westgate, and Ben Wisner. 1976. "Taking the Naturalness out of Natural Disasters." *Nature* 260 no. 5552 (April): 566–567.

Sauvy, Alfred. 1952. "Trois Mondes, Une Planète." *L'Observateur*, August 14: 1952. http://www.homme-moderne.org/societe/demo/sauvy/3mondes.html.

Standing, Guy. 2011. "The Precariat-The New Dangerous Class." *Policy Network Essay*, May 2011. https://eprints.soas.ac.uk/15711/1/Policy%20Network%20article,%2024.5.11.pdf

Standing, Guy. 2014. *The Precariat: The New Dangerous Class.* London: Bloomsbury.

Szerszynski, Bronislaw, and John Urry. 2010. "Changing Climates: Introduction." *Theory, Culture and Society* 27 no. 2–3: 1–8.

UN. 2020. *World Social Report 2020: Inequality in a Rapidly Changing World.* United Nations Publications. https://www.un.org/development/desa/dspd/wp-content/uploads/sites/22/2020/01/World-Social-Report-2020-FullReport.pdf.

UNDRR. 2020. *The Human Cost of Disasters: An Overview of the Last 20 Years (2000–2019).* UN Office for Disaster Risk Reduction. https://reliefweb.int/sites/reliefweb.int/files/resources/Human%20Cost%20of%20Disasters%202000-2019%20Report%20-%20UN%20Office%20for%20Disaster%20Risk%20Reduction.pdf.

UNHCR. 2020. *Mid-Year Trends 2020.* Copenhagen: United Nations High Commissioner for Refugees. https://www.unhcr.org/statistics/unhcrstats/5fc504d44/mid-year-trends-2020.html.

Watts, Michael. 1983. *Silent Violence: Food, Famine and Peasantry in Northern Nigeria.* Berkeley, California: University of California Press.

Watts, Michael. 1993. "Hunger, Famine and the Space of Vulnerability." *GeoJournal* 30 no. 2: 117–125.

White, Gregory. 2011. *Climate Change Migration: Security and Borders in a Warming World.* Oxford: Oxford University Press.

Wisner, Ben. 1993. "Disaster Vulnerability: Scale, Power and Daily Life." *GeoJournal* 30 no. 2: 127–140.

Wisner, Ben, Piers Blaikie, Terry Cannon, and Ian Davis. 2004. *At Risk: Natural Hazards, People's Vulnerability and Disasters.* London: Routledge.

Wisner, Ben, J.C. Gaillard, and Ilan Kelman eds. 2012. *The Routledge Handbook of Hazards and Disaster Risk Reduction.* London: Routledge

World Bank. 2015. *World Development Report 2015: Mind, Society, and Behaviour.* Washington, DC: World Bank. http://www.worldbank.org/en/publication/wdr2015

PART I

Why vulnerability still matters

2

REMAKING THE WORLD IN OUR OWN IMAGE

Vulnerability, resilience, and adaptation as historical discourses[1]

Greg Bankoff

For most of humanity, disasters have always been a 'frequent life experience' (Bankoff 2003). Now a warming climate and less predictable weather patterns, as well as an expanding urban infrastructure susceptible to geophysical hazards, make the world an increasingly dangerous place, even for those living in high-income countries (HICs). It is an opportune moment, therefore, from the vantage point of the third decade of the twenty-first century, to review the terms and concepts that have been employed regularly over the past 50 years to assess risk and to measure people's exposure to such events to determine whether or not they are still valid. In particular, it is useful to examine 'vulnerability', 'resilience', and 'adaptation', the principal theoretical concepts that, from a historical perspective, have dominated disaster studies since the end of the Second World War and enquire as to what extent they were discourses particular to their time and place.

That time and place was the Cold War in Europe, an ideological contest that sought to explain societies *and their environments* from the stance of competing conceptual frameworks and then, in its aftermath, the 'triumph' of liberal democracy, neoliberal economics, and, in recent decades, globalisation. The discourses elaborated to describe these decades all owe their origins to a Western intellectual tradition that cast the rest of the world as disease-ridden, poverty-stricken, and hazard-prone regions dependent on external medical knowledge, overseas aid, and scientific expertise (Bankoff 2001). During the Cold War, the non-Western world was depicted as vulnerable, and then following the collapse of the Soviet Union, as resilient. More recently, the focus has been more on climate change through policies that advocate adaptation and disaster risk reduction (DRR) as the guiding principles of disaster risk management (DRM). Although all these discourses have been present in one form or another throughout most of this period, there has been more recently something of an intellectual adjustment that has rendered vulnerability seemingly less important as a discourse.

DOI: 10.4324/9781003219453-3

If vulnerability helped to elucidate how the world was rendered unsafe in the second-half of the twentieth century, why has the term lost favour, at least on an official level, at the turn of the twenty-first century when the environmental and societal conditions that inspired its formulation are, if anything, more prevalent?

Vulnerability as a Cold War discourse

The historical context was highly significant to the emergence of vulnerability as a discourse. The term emerged and gained validity during the 1970s, a time when the Cold War was heating up again. Its chief proponents were practitioners and scholars motivated by concern with the plight of citizens in the newly denominated Third World, and who shared a growing suspicion of the development policies pursued by Western governments and transnational corporations in these new nations. The Cold War entrenched a militarised model of civil defence in the years following the Second World War, which subsumed disaster management under the need for nuclear preparedness. The North Atlantic Treaty Organisation, for instance, established a Civil Defence Committee in 1951 to provide protection for its citizens, stating that 'the capabilities to protect our populations against the effects of war could also be used to protect them against the effects of disasters' (NATO 2001, 5). However, by demonstrating that there was nothing 'natural' about natural disasters and that people were put at risk as much by the political and social structures of the societies in which they lived as by any physical event or hazard, some scholars began to question the hitherto unchallenged assumption that the growing incidence of disasters was due to a rising number of purely natural phenomena. In the process, they offered a searing critique of both the means and the intent behind Western-led development and investment policies (O'Keefe, Westgate and Wisner 1976; Hewitt 1983; Watts 1993). Rather than lifting people out of poverty, too often the outcomes of such programmes were to make of their lives a 'permanent emergency' (Wisner 1993, 131–133). The emphasis, instead, was shifted from an agent-specific focus on an extreme event to consideration of what rendered communities unsafe, a condition that depended primarily on a society's social order and the relative position of advantage or disadvantage that a particular group occupied within it (Cannon 1994; Hewitt 1997, 141). The term coined to evaluate the nature and extent of this risk was 'vulnerability', a measure not only of people's exposure to hazards but also of their capacity to recover from loss (Chambers 1989; Blaikie et al. 1994; Hewitt 1997; Lewis 1999; Cannon 2000; Pelling 2003; Bankoff, Frerks and Hilhorst 2004; Adger 2006).

The purpose here is not to assess the merits of vulnerability as a term in relation to any other, but simply to examine it historically as a product of its time and place, and the importance of the relative economic and political factors that underlay its conceptualisation (Cote and Nightingale 2012, 478). The Cold War origins of the term are to be found in its definition or, rather, the way in which vulnerability is applied in practice. Everybody, of course, is made vulnerable to some extent by a combination of variables such as age, class, disability, ethnicity, and gender, among others, affecting their entitlement to command basic necessities and their empowerment to enjoy fundamental rights (Watts 1993, 118–120). While the term embraces a wide spectrum of who is vulnerable, in practice, the spotlight is primarily on those with the highest degree of constant exposure to risk. Furthermore, these people overwhelmingly live in low- and middle-income countries (LMICs). The relative vulnerability of these populations is usually defined either in terms of mortality or magnitude, such as the Bhola Cyclone of 1970 that killed an estimated 500,000 people in East Pakistan, now Bangladesh (Sommer and Mosley 1972), or the 7.8 magnitude Tangshan earthquake of 1976 that flattened a city in northeast China, claiming the lives of approximately 250,000 people (Yong 1988). Vulnerable people, it was apparent, lived in vulnerable places, and these vulnerable places were principally situated in the so-called developing world and were subject to the monolithic industrial modernisation projects of the post-Second World War era.

This message was made clear in the most complete model proposed to clarify how risk is generated and disasters materialise. In *At Risk: Natural Hazards, People's Vulnerability and Disasters*, Piers Blaikie et al. (1994) presented the pseudo-formula *risk = hazard + vulnerability* to demonstrate how the measure of a community's risk is directly attributable not only to the physical hazard experienced but also to the extent to which a particular social order puts people at risk. According to the Pressure and Release (PAR) model, vulnerability is reproduced over time: at a global level through 'root causes' that reflect the historical distribution and exercise of power in a society that marginalises certain groups; at an intermediate level through more contemporary 'dynamic pressures' that include conflict, epidemic disease, foreign debt, urbanisation, and certain economic policies and environmental degradation; and at an immediate local level through 'unsafe conditions' that equate to a particular group's hazardous living conditions, dangerous livelihoods, or inadequate food sources (Blaikie et al. 1994). At the same time as offering a framework for linking the impact of hazards to a series of societal factors and processes that generate vulnerability, the PAR model exposed the processes that transformed the colonial territories of the post-Second World War into the new states of the Third World.[2] The critique was unequivocal: imperial heritage, development policies, and unequal power relationships rendered some communities less able to deal with disasters and left them more at risk.

Not that everybody was affected in the same way or to the same extent. A small proportion of households and enterprises, more in wealthier states, did benefit from development policies and were able to protect their families and assets from the worst of human and natural excesses. To paraphrase the presidential campaign slogan of US Democratic Party hopeful Adlai Stevenson in 1952, some people were safer than at any time in history and had 'never had it so good'. A growing class of middle-income earners who were well-educated and politically engaged—middle managers, professionals, technicians, and even unionised workers—were also relatively safer, even if subject to the economic vagaries of globalisation. These groups, however, never constituted more than a small minority of the world's population. The rest, the majority of humanity, whose lives were overwhelmingly rendered vulnerable and whose deaths mainly constituted the figures in the newly compiled disaster statistics, comprised the low-income populations of the Third World. These people wielded little political influence and had fewer entitlements. They also included a persistent if fluid section of First World citizens whose lives were rendered insecure by a combination of class, ethnicity, gender, or some other factor.

Vulnerability offered a means of critiquing developmentalism and the untrammelled pursuit of material prosperity that had become the dominant model of economic progress after 1945. Arturo Escobar (1995a, 213) refers to this conceptual ascendancy as 'colonisation', indelibly shaping representations of reality and constructing 'the contemporary Third World, silently, without our noticing it'. Nations were increasingly assessed in terms of development or lack of it and some societies began to be regarded (and regard themselves) as underdeveloped, a condition viewed as synonymous with backwardness, poverty and, implicitly, vulnerability (Escobar 1995b, 5). The Third World was not only disease-ridden and poverty-stricken, but it was also increasingly disaster-prone, a zone where repeated hazards inflicted upon people sudden death and damaging losses that left communities physically weak, economically impoverished, socially dependent, and psychologically harmed. It also formed an integral part of a generalising, Western cultural discourse that denigrated large regions of world as dangerous (Bankoff 2001).

Development was supposed to ameliorate the unsafe conditions and dynamic pressures that put people at risk. If it largely failed to do so, it was because development was too much a part of the root causes that underlay societies' vulnerability in the first place. In this newly constructed Third World, many people began to perceive development projects, which required the conversion of prime agricultural or seafront land to industrial and commercial usage, as disadvantageous rather than beneficial (Heijmans 2004). To make way for such projects, local communities frequently were displaced without consultation, losing not only their homes, livelihoods, and rights to cultivate land, but also their dignity, identity, and roots. Moreover, the increasing dependence of industrialised societies on fossil fuels—more than one-half of the total oil consumed in the past 150 years has been burnt in the past three decades—necessitated an ever-increasing expanse of land and the organisation of a vast workforce outside of HICs to supply its need for all forms of energy (Mitchell 2011, 6, 16). It is hardly surprising that many environmentalist and grassroots activists in these affected countries began to talk about 'development aggression', a form of development in which people were neither the partners nor the beneficiaries of projects, but rather its victims (Heijmans 2004). It was also a condition that rendered societies and their environments much more vulnerable to the effects of natural hazards.

The battle over resilience and the rise of neoliberalism

The link between development and disasters, the Cold War and vulnerability was not immediately apparent. Indeed, it is not a connection that is often made even today. However, with the end of the Cold War in 1991—and, incidentally, the demise of the Third World, at least in name—the emphasis on how societies should be viewed began to shift. Gradually, it was suggested that the issue of vulnerability should be turned around and approached from a more positive viewpoint. Societies were seen no longer as simply vulnerable, with all its associated negative connotations, but people began to be considered as primarily resilient with capacities to organise, resist, learn, change, and adapt (Handmer 2003). This change in thinking was already well under way, influenced by the work of the ecologist Crawford Holling. Holling (1973) maintained that ecosystem dynamics were best understood in terms of a system's capacity to absorb disturbance while retaining the same population or state variables. These same forces, it was argued, were also at work in a social context. First gaining official approbation during the oil crisis of the 1970s, resilience thinking moved away from a qualitative assessment of why people were at risk towards a consideration of the available response options (Walker and Cooper 2011, 153).

A change of discourse was also politically expedient in the new international climate. The rationale behind Overseas Development Aid (ODA) initiated by US President Harry Truman in 1949 and projects funded by the World Bank designed to contain and roll back the spread of communism was no longer required. Development, in this sense, had been a continuation of a colonial discourse refined in the debates regarding post-war compensation about the best ways to deal with poverty, frequently compounded by natural hazards and disasters. As the anti-communist agenda receded in the 1990s, structural adjustment loans, foreign direct investment, and private capital flows began to replace ODA as the favoured development paradigm. Any debate about the relative merits of market-oriented reform simply 'expired' (Summers and Pritchett 1993, 385). At the heart of the new approach was a neoliberal or strongly market-based view of promoting growth through fiscal adjustments followed by facilitating macroeconomic stability and integration into the international economy (Easterly 2005). If anything, this neoliberalism was a throwback, at least in principle, to the nineteenth century in its heavy reliance on free-market mechanisms. Under the new financial regime, funding was made conditional on deregulation, fiscal discipline, privatisation, a reduced role for the state, tax reform, and trade liberalisation (Veltmeyer 2005). The consequent privatisation of public services and infrastructure and the selling off of state assets commonly occurred in the absence of proper regulatory safeguards, placing many services beyond the reach of the poor, leaving others at the mercy of substantial rises in utility charges, and rendering them all more vulnerable to the impacts of natural hazards and disasters (Hilary 2004).

In this new political climate, it was expedient to stress what made people resilient rather than what made them vulnerable (Maskrey 1989; Blaikie et al.

1994; Folke 2006; Manyena 2006; Gaillard 2007; Alexander 2013). Resilience was regularly referred to in terms of a community's social capital or the manner in which a contribution freely given was expected to be reciprocated at an appropriate time, and by the development of group relations that morally enforced this code (Woolcock 2001). The role of social capital in disaster management has received increasing, if not uncritical attention (Fine 2010), in recent years, both with regard to volunteerism in the aftermath of major events such as the Kobe earthquake of 1995 or the Marmara earthquake of 1999, and to everyday community risks (Jalali 2002; Nakagawa and Shaw 2004; Bankoff 2015). Emphasis has also been placed on the importance of location in generating particular forms of associational activities, a geography of social capital, and 'a recognition that context matters to the outcomes of social processes' (Mohan and Mohan 2002, 202).

If vulnerability was a product of the Cold War and the conceptual framework that created the Third World, to what extent is resilience an 'invention' of a way of thought that promotes and condones neoliberalism? The uncomfortable truth is that the two discourses have much in common and share many policy approaches, even if for different reasons (Walker and Cooper 2011; MacKinnon and Derickson 2012). The neoliberal agenda envisages a state where human well-being is best advanced by 'the maximisation of entrepreneurial freedoms within an institutional framework characterised by private property rights, individual liberty, unencumbered markets, and free trade' (Harvey 2007, 22). To achieve these desired ends, the state only has one primary responsibility: to create the conditions that permit a fully-functioning market. In this 'voluntary state', where the emphasis has shifted from the structural factors that cause vulnerability to individual responsibility and choice, all other responsibilities are labelled as 'personal'. A resilient community is one that is 'better able to weather its exposure to global financial markets through the adoption of a localised, decentralised, post-carbon, ecosystems-based model of growth' (Walker and Cooper 2011, 155). In effect, the state devolves public safety to civil society and then expects the market to meet the social needs of the population. It does so by promoting the conditions that create wealth and then allowing the wealthy to volunteer assistance to those it has impoverished. This 'hollowing out' of the state, however, cannot be achieved without the voluntary contribution of non-state actors.

In relation to disaster management, the state increasingly depended on non-governmental organisations (NGOs) to fulfil the public safety roles of which it wished to be divested, if not in the immediate short term in respect to the provision of emergency services, then certainly in the longer term as regards preparedness, mitigation, recovery, and reconstruction. From the neoliberal perspective, divesting humanitarian assistance to NGOs was a salutary alternative to funding corrupt governments in LMICs. Reframing the state's responsibilities in this manner now cast poverty as largely a voluntary choice: the poor chose to be poor and only had themselves to blame for being poor. Likewise, those who were vulnerable chose to be vulnerable and had only themselves to blame for being vulnerable (Nickel and Eikenberry 2007, 536–537). Echoing the harsh sentence of nineteenth-century Social Darwinists, proponents of neoliberalism viewed social responsibility as optional, and vulnerability as voluntary. 'Resilient' people do not have to look to the state to secure their well-being as they have made themselves secure already. 'Social resilience' has become a core constituent of the neoliberal economic agenda now expressed in terms of sustainable development and its prescriptions for institutional reform: '[r]esilience was reconceived not simply as a property of the biosphere, in need of protection from the economic development of humanity, but a property within human populations which now needed promoting through the increase of their "economic options"' (Reid 2012, 72).

The commonalities in practice between a neoliberal agenda and the shift from vulnerability to social resilience in DRM brought to the fore a new rhetoric that emphasised DRR and focused on community-based disaster risk management (CBDRM). DRR or a mitigation approach began to emerge in the 1970s and gradually became the most dynamic discourse in the global policy field of disasters (Hannigan 2012, 130–145). If resilience recognised the necessity of incorporating ecological systems thinking into disaster management through a greater awareness of environmental and sustainable development issues, the priority of DRR was risk reduction and prevention through improving livelihoods and increasing social mobilisation. Pre-disaster mitigation was first piloted in the US in 1997 (Project Impact) and passed into law in 2000 (Disaster Mitigation Act of 2000) (McCarthy and Keegan 2009). Its international ascendancy can be noted in the five priorities identified in the Hyogo Framework for Action 2005–2015 (UN 2005), and, more recently, in the four priorities agreed upon in the Sendai Framework for Disaster Risk Reduction 2015–2030, which formally recognised the responsibility of local government, the private sector, and other stakeholders alongside that of the state in reducing disaster risks (UNISDR 2015).[3]

CBDRR offered the means to translate DRR into action. It claims an effective and sustainable approach to disaster reduction by empowering people to tackle the underlying problems of poverty, marginalisation, environmental degradation, and political abuse (World Bank 2001). CBDRR is distinguished by emphasising participatory processes in disaster management, capacity-building, removal of the root causes of vulnerability, and the mobilisation of the less vulnerable sectors in support of those with needs (Heijmans and Victoria 2001, 13–18). In the Philippines, for example, NGOs have become increasingly involved in DRM and have integrated mitigation and preparedness into their existing operations. At the same time, existing development organisations have expanded their programmes to incorporate disaster management capabilities, and specific NGOs have formed in direct response to actual disasters to conduct integrated relief and rehabilitation work (Luna 2001, 219–220). In this sense, DRR is only rediscovering that people with local knowledge and expertise are the principal resource of their community. As Andrew Maskrey (1989, 87) succinctly observed more than 25 years ago, 'only local people know their own needs and therefore only they can define their own priorities for mitigation, within a given context'.

Participatory approaches to DRR, though, have not always proven to be the panacea they were once hoped to be and have been suborned by the World Bank and other multilateral lending institutions to serve a neoliberal agenda. Initially, participatory approaches were held up as a major counterbalance to the power of the dominant development discourse (Chambers 1997). Yet, as participation was increasingly written into development projects, it has become 'wholly compatible with the liberalisation agenda, and poor people's voices carefully marshalled to provide support for the Bank's policy prescriptions' (Williams 2004, 558). Instead of local knowledge shaping development projects, they are in fact often shaped more by locally dominant groups and by the project's own interests. Rather than people participating in agency programmes, it is the other way round to ensure consistency with project-defined models; what David Mosse (2001, 24) calls the 'ventriloquization' of villagers' needs. That is, participatory approaches serve the dual purpose of depoliticising the question of poverty and shifting responsibility for the project's success from the administrating agency to the participants (Williams 2004, 564–565). Even if CBDRR acknowledges the need for local participation in disasters, it is communities that ultimately are responsible for improving their capacity and addressing the risks.

By different routes and for very different intentions, neoliberalism and social resilience end up advocating much the same approach by much the same methods. Both emphasise an active citizenship whereby people take responsibility for their own social and economic well-being, and both share a general distrust of centralised state systems and a desire to decentralise responsibilities. The stress is on local capacity, local decision-making, local responsibility, and, of course, local funding. To one, however, this championing of civil society is a way to disguise the imposition of market discipline, part of a state-building agenda, which, far from empowering people, is a means of exercising 'governance from a distance' (Joseph 2013). To others, however, resilience is a continuing critique of existing international development and aid, which, far from shedding its Cold War agenda, only found new vitality in the policies and programmes associated with the Washington Consensus. The continuing notion that 'natural disasters' are simply a sign of underdevelopment and that the poor suffer disproportionately during such events because of their underdevelopment was bitterly attacked. In the wake of Hurricane Mitch in 1998, for instance, trust in this principle was used to argue that economic development was the best answer to disasters in Nicaragua (Rocha and Christoplos 2001, 246). Ben Wisner (2001) decried the 'phantom decentralisation' in neighbouring El Salvador whereby government responsibilities were decentralised to local agencies without the funding or resources to implement them. Despite the encouraging rhetoric, the government's post-Mitch recovery plan 'produced vulnerabilities that affect all but the very richest' and was nothing more than 'run-away capitalism justified by neoliberal ideology' (Wisner 2001, 261). Naomi Klein (2007) has gone further and claimed that neoliberalism, even if it does not promote disasters, certainly profits from them, a process she aptly names 'disaster capitalism'.

Resilience is no less a Western discourse than is vulnerability: it recasts the world according to culturally specific dictates. Depending on the context in which it is evoked, resilience either tries to restructure non-Western societies according to prescribed economic formulae or it looks for salvation in the social structures of traditional communities that it defines to its own intent. All too often, however, it is a profoundly conservative discourse. States promote resilience to mask inequalities and social differentiation within societies, absolving themselves of their duty of care and implicitly accepting capitalism as an immutable force akin to the power of nature (MacKinnon and Derickson 2012, 258). As a critique of the status quo, on the other hand, CBDRM exalts existing social relations within communities and denies the state a legitimate role in promoting change in society. In either case, it is, as Cote and Nightingale (2012, 484–485) suggest, 'a power-laden framing that creates certain windows of visibility on the processes of change, while obscuring others'.

Globalisation and the 'turn' to Adaptation

The all but disappearance of communism and the rise of neoliberalism in the last quarter of the twentieth century prepared the way for the integration of economies, industries, markets, and cultures on a truly global scale. This new network society is both informational and global (Castells 1996). The process of globalisation, of course, has been taking place for centuries, but it has speeded up and diversified enormously over the past 50 years. Previously confined to mainly economic matters, the term now includes activities such as technology, media, and culture. Globalisation is the ultimate realisation of neoliberal thinking expressed on a planetary scale, and, as Klein (2015) writes, is an ideological project 'to lock in a global policy framework that provided maximum freedom to multinational corporations to produce their goods as cheaply as possible and sell them with as few regulations as possible – while paying as little in tax as possible' (Klein 2015, 19).

In this new political era, the existential threat is no longer Reds under the beds or the public sector of nation states but is climate change (Leichenko and O'Brien 2008). Scientific evidence now supports the conclusion that human activity is changing the climate and will continue to affect it for years to come, even if there are no further emissions of greenhouse gases. World temperatures have fluctuated in the past. However, at no time in the past 650,000–800,000 years have concentrations of carbon dioxide, chlorofluorocarbons, hydrofluorocarbons, water vapour, methane, nitrous oxide, and ozone in the atmosphere been so high as they are today (Giddens 2014, 12; World Bank 2015, 16). While sceptics remain (Lomborg 2001; Bell 2011), the current debate is more about the nature of climate change: whether it is a gradual process that will allow human societies to adjust to the new conditions, or whether it is non-linear, whereby crossing some threshold will precipitate sudden and catastrophic change (Lovelock 2006). There is little that can be done in the latter case apart from ensuring that this tipping point is not reached by reducing emissions into the atmosphere. Alternatively, if climate change is relatively slow, then societies will have time to adapt given the necessary inducements and incentives (Szerszynski and Urry 2010, 1–2). The key concept in this new discourse is climate change adaptation.

Adaptation as a concept, however, is also a contested domain. This recent turn in discourse heralds yet another conceptual power struggle between Western governments, financial institutions, multinationals, and LMICs over how to shape the future (Pelling 2011, 3). Unlike vulnerability and resilience, though, adaptation is very much a top-down rather than a bottom-up concept largely conceived and implemented by the UN and international organisations. Its definition and application are fought over in much the same way as were vulnerability and resilience. As the increase in greenhouse gas emissions began to be taken seriously by governments and scientists, an international treaty was signed in 1992, the United Nations Framework Convention on Climate Change (UNFCCC), in which the climate was acknowledged as 'a common concern of humankind' (UN 1992). The UN established the Intergovernmental Panel on Climate Change (IPCC) to provide an objective, scientific view of climate change, its political and economic impacts, and the options available for mitigation and adaptation. Tellingly, the IPCC describes adaptation in terms of an 'adjustment in natural or human systems in response to actual or expected climatic stimuli or their effects, which moderates harm or exploits beneficial opportunities' (IPCC 2007, 809).

As so stated, adaptation is defined as an inherently conservative activity that functions to preserve the status quo rather than to encourage more radical solutions that might threaten existing social and political systems. This is hardly surprising given its provenance in a UN system beholden to the nation states that fund its institutions and agencies, and the banking and corporate interests that manipulate the policies and interests of national governments (Pelling 2011, 11). Adaptation's preoccupation with climate science, whether the intensity of storms will mount, how high sea levels will rise, and to what extent floods will become more frequent, runs the risk of blaming nature once more for disasters and returning to an older hazard-focused paradigm that ignores how such events are socially and physically constructed. This is a familiar trope that renders large parts of the world as vulnerable by blaming the poverty of these regions squarely on natural forces and disasters on people's lack of resilience. It is part of the conceptual vocabulary of neoliberalism that evokes a Social Darwinist ethic implying that those who do not adapt are not fit to survive. 'It burdens and blames the victim', according to Jesse Ribot (2011, 1160), 'by devolving the onus of adjustment to the organism or affected unit'. It also serves to diffuse the opprobrium that might otherwise be directed at an economic system created by, and largely still benefiting, Western industrialised nations, by stressing the common plight of humanity and by concentrating discussions on purely climatic and scientific phenomena (Bankoff 2001).

There is an unwillingness to recognise that the overdevelopment of the industrialised West has been at the expense of the underdevelopment of the rest of the world whose peoples are now expected to bear an inordinate share of the socio-environmental consequences of climate change. In 2013, a group of Pacific island nations came close to asking the International Court of Justice for an advisory opinion on who was responsible for global warming. They only refrained from pursuing a claim because they were advised to wait until the science made for more irrefutable evidence provided by the Fifth Assessment Report of the IPCC in 2014. In the same year, 132 LMICs staged a walk-out at the UN's Climate Change Conference in Warsaw, Poland, in protest at the attempt by Western countries to block all talk about compensation for the impacts of global warming until after 2015 (Weymouth 2013). Evoking Cold War rhetoric, Australian Prime Minister Tony Abbott referred to his country's carbon tax, imposed by a previous left-of-centre government, as 'basically socialism masquerading as environmentalism' (Vidal 2013). Rather than being a unifying issue around which humanity rallies in the face of a common challenge, climate change is as divisive as any Cold War discourse, although the geographical fracture lines are now drawn as North–South rather than East–West. The threat is also global, even if it disproportionately affects more equatorial regions.

Moreover, climate change frequently serves as a scapegoat to explain the causes of natural hazards and occlude the true nature of disaster (Kelman and Gaillard 2008; Mercer 2010). While climate change is an important driver of certain types of hazards, especially hydrological ones, it is often used as an excuse to focus on natural explanations rather than social ones (Kelman et al. 2016). This is what Ribot (2014) describes as drawing attention to the *who* are vulnerable rather than the *why* they are vulnerable question; the latter is too socially and politically contentious to address. Most government agencies and development organisations invested with climate policy prefer to maintain existing structures and relationships: adaptation is conceived and implemented in such a manner that most projects preserve rather than challenge the status quo (Pelling, O'Brien and Matyas 2015). Based on published research, the IPCC's reports are largely 'a product of negotiated content between science and governments' and rarely risk alienating the political and technical decision-makers on whose support they depend (Pelling 2011, 37–38). Consequently, as disasters are attributed solely to climate change, global institutions, following the UNFCCC's lead, craft adaptation funds to redress only the 'additional' damages produced in this manner, limiting liability and avoiding all consideration of the root causes of what made people vulnerable to climatic variations in the first place (Ribot 2014, 670–672). The talk once again is about what makes people resilient and the discourse on adaptation is centred on how to maintain what existed before. Such attitudes remain prevalent among institutions such as the Overseas Development Institute and the United Nations Office for Disaster Risk Reduction (Kelman et al. 2016, S133).

Of course, adaptation need not be depicted in this manner and the threat of climate change can raise profound questions about existing paradigms of development (Godfrey-Wood and Naess 2016). Mark Pelling (2011) identifies three levels at which adaptation can influence development: (i) adaptation to build resilience through implementing changes that do not question the underlying assumptions or power asymmetries in society; (ii) a transitional stage of adaptation that encourages only incremental changes in rights and responsibilities without advocating a fundamentally different regime; and (iii) transformational adaptation that advocates radical reform of the economic and political systems and the cultural discourses on which they are based. While he is careful not to favour any one form of adaptation, it is clear that the challenges posed by climate change demand more radical solutions than simply the resilience favoured by the IPCC and the UNFCCC. Such advocacy is anathema to hard-core conservatives who prefer to deny that climate change is even real, owing to fear of opening the door once again to massive state intervention and market regulation (Klein 2015, 40).

Conclusion

During the Cold War, vulnerability offered a needed critique of development policies that emphasised growth rather than 'purposeful development' (Cannon and Müller-Mahn 2010, 623–626). After 1991, the stress placed on resilience signified a shift away from the extent to which socioeconomic systems exposed people to different levels of risk to a perspective that underscored how human actions made it possible for social-ecological systems to survive. Vulnerability remains a significant consideration in these discussions if for no other reason than it has an adverse effect on social-ecological resilience (Adger 2006, 269). The neo-liberalism that dominated the decades following the collapse of the USSR suborned much of the public and academic discourses surrounding resilience by championing individual choice and personal responsibility. In the process, vulnerability was rendered an almost voluntary condition, one that was mainly the result of poor individual decisions. Adaptation, as it is presently conceived and implemented through the IPCC and the UNFCCC, is shown to be little more than a form of resilience in another guise, although, like the latter, it too has the potential to be a conduit for more radical change. The present focus on climate change and the need for social adaptation runs the risk of reducing the latter to a choice freely made by individuals, communities, and states. As Cannon and Müller-Mahn (2010, 627) deftly point out, for many, being 'risk adverse' in the present is actually nothing more than a neoliberal concern with 'profit maximisation' in the longer term. How societies best adapt to climate change is effectively reduced to a question of how far growth can continue while limiting the most serious environmental ramifications and even profiting from the economic opportunities that arise. Nor is there any guarantee that people made fully aware of the perils of climate change will respond by adopting risk reduction measures and behaviours.

All three discourses are tainted in one respect: they are all culturally specific to Western perspectives. Both vulnerability and resilience are discourses that originated at a particular historical juncture. Their meanings were shaped by a particular historical perspective and their significance can only be understood through consideration of the way power operated at the time in the prevailing socio-environmental systems. If vulnerability expressed a profound unease with the developmental model that dominated the Cold War era and that depicted natural hazards as largely physical events for which there were mostly technical solutions, the subsequent discourse of resilience fitted well with neoliberal ideas about competition and entrepreneurship that viewed disasters after the collapse of communism as principally the outcome of individual choice. The current emphasis on adaptation as a trope implies accepting a world in which disturbance and crisis are constant features whether caused by climate change and/or social upheaval. It is also one where there is a continual need for neoliberally sanctioned discourses about resilience and change. It accepts disaster as an endemic condition in anticipation of which society must remain on a permanent state of high alert. It is also a profoundly conservative discourse that largely obscures questions about the role of power and culture in society, and about whose environments and livelihoods are to be protected and why (Cote and Nightingale 2012, 484–485; Krüger et al. 2015).

There seems no escape from remaking the world again and again after a particular cultural image. No rival discourse seems ready yet to challenge Western hegemony in the language and metaphor of international governance and development policy. Accordingly, disasters remain inherently political events 'because they pose questions about who should be allowed to re-compose the world and how' (Guggenheim 2014, 4). That these discourses are primarily conservative and inherently protect Western interests is not unexpected given the historical context in which they evolved. Only vulnerability offers a critique of existing power relations and the status quo, but its import was blunted by the end of the Cold War and the new focus on resilience. Adaptation, too, has not so much been subverted by a neoliberal agenda as it has been chiefly conceived in its likeness and has been mainly implemented by its instruments and agencies. If the stress in HICs is increasingly on the need for adaptation and necessary adjustment as the only really practical measure, what, in effect, makes societies more resilient, for those in LMICs, the issue still remains much more about what renders them vulnerable, more especially as that condition is seen as largely imposed by the West on the rest (Bankoff and Borrinaga 2016). 'Rather than seeking causality in social history', Ribot (2014, 671) concludes, 'adaptation becomes a necessary adjustment to the droughts, floods or storms that are directly attributable to climatic events'.

Examining these dominant discourses as products of the historical forces that gave them birth exposes the underlying values and norms that continue to shape the world and how one chooses to frame the future. Unfortunately, the power relations that underlie these discourses have not changed significantly since the Second World War, as the world has largely been remade again and again according to an image fashioned by certain sectors in Western societies. Only through the placing of continuing emphasis on the root causes that make people vulnerable, on how power relations operate in society to place some people more at risk than others, on the importance of culture to community resilience, and on how adaptation provides an opportunity for a radical change in the way in which human societies operate can a similar fate be prevented and history made to stop repeating itself—yet again.

Notes

1 A version of this article first appeared in *Disasters* (2019. 43 no. 2: 221–239).
2 Terms such as 'Third World', 'First World', 'less developed', 'developing', and 'developed' are employed in this chapter where they are appropriate to the historical context in which they were used.
3 The five priorities of the Hyogo Framework for Action are: ensuring that DRR is a national and a local priority; identifying, assessing, and monitoring disaster risks and enhancing early warning; using knowledge, innovation, and education to build a culture of safety and resilience; reducing the underlying risk factors; and strengthening disaster preparedness for effective response. The four priorities for action of the Sendai Framework for Disaster Risk Reduction are: understanding disaster risk; strengthening disaster risk governance; investing in disaster risk reduction for resilience; and enhancing disaster preparedness for effective response.

References

Adger, W. Neal. 2006. "Vulnerability." *Global Environmental Change* 16 no. 33: 268–281.

Alexander, David. 2013. "Resilience and Disaster Risk Reduction: An Etymological Journey." *Natural Hazards and Earth Systems Science* 13 no. 11: 2707–2716.

Bankoff, Greg. 2001 "Rendering the World Unsafe: 'Vulnerability' As Western Discourse." *Disasters* 25 no. 1: 19–35.

Bankoff, Greg. 2003. *Cultures of Disaster: Society and Natural Hazard in the Philippines*. London: Routledge.

Bankoff, Greg. 2015 "'Lahat para sa Lahat' (Everything to Everybody): Consensual Leadership, Social Capital and Disaster Risk Reduction in a Filipino Community." *Disaster Prevention and Management* 24 no. 4: 430–447.

Bankoff, Greg, and George E. Borrinaga. 2016. "Whethering the Storm: The Twin Natures of Typhoons Haiyan and Yolanda." In *Contextualising Disaster*, edited by Gregory V. Button and Mark Schuller, 44–65. New York: Berghahn Books.

Bankoff, Greg, Georg Frerks, and Dorothea Hilhorst, eds. 2004. *Mapping Vulnerability: Disasters, Development, and People*. London: Earthscan.

Bell, Larry. 2011. *Climate of Corruption: Politics and Power behind the Global Warming Hoax*. Austin, TX: Greenleaf Book Group.

Blaikie, Piers, Terry Cannon, Ian Davis, and Ben Wisner. 1994. *At Risk: Natural Hazards, People's Vulnerability and Disasters*. London: Routledge.

Cannon, Terry. 1994. "Vulnerability Analysis and the Explanation of 'Natural Disasters'." In *Disasters, Development and Environment*, edited by Ann Varley, 13–30. Chichester: John Wiley and Sons.

Cannon, Terry. 2000. "Vulnerability Analysis and Disasters." In *Floods*, edited by Dennis Parker, vol. 1 45–55. London: Routledge.

Cannon, Terry, and Detlef Müller-Mahn. 2010. "Vulnerability, Resilience and Development Discourses in Context of Climate Change." *Natural Hazards* 55 no. 3: 621–635.

Castells, Manuel. 1996. *The Rise of the Network Society.* Oxford: Blackwell Publishers Limited.

Chambers, Robert. 1989. "Vulnerability, Coping and Policy." *IDS Bulletin* 20 no. 2: 1–7.

Chambers, Robert. 1997. *Whose Reality Counts? Putting the First Last.* London: Intermediate Technology Publications.

Cote, Muriel, and Andrea J. Nightingale. 2012. "Resilience Thinking Meets Social Theory: Situating Social Change in Socio-Ecological Systems (SES) Research." *Progress in Human Geography* 36 no. 4: 475–489.

Easterly, William. 2005. "What Did Structural Adjustment Adjust? The Association of Policies and Growth is Repeated IMF and World Bank Adjustment Loans." *Journal of Development* Economics 76 no. 1: 1–22.

Escobar, Arturo. 1995a. "Imagining a Post-Development Era." In *Power of Development*, edited by Jonathan Crush, 211–227. London: Routledge.

Escobar, Arturo. 1995b. *Encountering Development: The Making and Unmaking of the Third World.* Princeton, NJ: Princeton University Press.

Fine, Ben. 2010. *Theories of Social Capital: Researchers Behaving Badly.* New York, NY: Pluto Press.

Folke, Carl. 2006. "Resilience: The Emergence of a Perspective for Social Ecological Systems Analysis." *Global Environmental Change* 16 no. 3: 253–267.

Gaillard, Jean-Christophe. 2007. "Resilience of Traditional Societies and Facing Natural Hazards." *Disaster Prevention and Management* 16 no. 4: 522–544.

Giddens, Anthony. 2014. *The Politics of Climate Change.* Cambridge: Polity Press.

Godfrey-Wood, Rachel, and Lars O. Naess. 2016. "Adapting to Climate Change: Transforming Development?." *Development Studies – Past Present and Future* 47 no. 2: 1–9.

Guggenheim, Michael. 2014. "Introduction: Disasters as Politics – Politics as Disasters." *The Sociological Review* 62 no. S1: 1–16.

Handmer, John. 2003. "We Are All Vulnerable." *Australian Journal of Emergency Management* 18 no. 3: 55–60.

Hannigan, John. 2012. *Disasters without Borders.* Cambridge: Polity Press.

Harvey, David. 2007. "Neoliberalism As Creative Destruction." *The Annals of the American Academy of Political and Social Science* 610 no. 1: 22–44.

Heijmans, Annelies. 2004. "From Vulnerability to Empowerment." In *Mapping Vulnerability: Disasters, Development and People*, edited by Greg Bankoff, Georg Frerks, and Dorothea Hilhorst, 115–127. London: Earthscan.

Heijmans, Annelies, and Lorna Victoria. 2001. *Citizenry-based and Development-oriented Disaster Response: Experiences and Practices in Disaster Management of the Citizens' Disaster Response Network in the Philippines.* Quezon City: Center for Disaster Preparedness.

Hewitt, Kenneth. 1997. *Regions of Risk: A Geographical Introduction to Disasters.* Edinburgh: Longman.

Hewitt, Kenneth, ed. 1983. *Interpretations of Calamity from the Viewpoint of Human Ecology.* London: Allen and Unwin.

Hilary, John. 2004. *Profiting from Poverty: Privatisation Consultants, DFID and the Public Services.* London: War on Want.

Holling, Crawford S. 1973. "Resilience and Stability of Ecological Systems." *Annual Review of Ecology and Systematics* 4 no. 1: 1–24.

Intergovernmental Panel on Climate Change. 2007. *Climate Change 2007 Synthesis Report.* Geneva: IPCC.

Intergovernmental Panel on Climate Change. 2014. *Climate Change 2014: Synthesis Report.* Contribution of Working Groups I, II and III to the Fifth Assessment Report of the IPCC. Geneva: IPCC.

Jalali, Rita. 2002. "Civil Society and State: Turkey after the Earthquake." *Disasters* 26 no. 2: 120–139.

Joseph, Jonathan. 2013. "Resilience As Embedded Neoliberalism: A Governmentality Approach." *Resilience: International Policies, Practices and Discourses* 1 no. 1: 38–52.

Kelman, Ilan, and Jean-Christophe Gaillard. 2008. "Placing Climate Change within Disaster Risk Reduction." *Disaster Advances* 1 no. 3: 3–5.

Kelman, Ilan, Jean-Christophe Gaillard, James Lewis, and Jessica Mercer. 2016. "Learning from the History of Disaster Vulnerability and Resilience Research and Practice for Climate Change." *Natural Hazards* 82 no. S1: S129–S143.

Klein, Naomi. 2007. *The Shock Doctrine: The Rise of Disaster Capitalism.* New York, NY: Metropolitan Books.

Klein, Naomi. 2015. *This Changes Everything: Capitalism vs. the Climate.* St Ives: Penguin Random House UK.

Krüger, Fred, Greg Bankoff, Terry Cannon, Benedikt Orlowski, and E. Lisa Schipper, eds. 2015. *Cultures and Disasters: Understanding Cultural Framings in Disaster Risk Reduction.* New York, NY: Routledge.

Leichenko, Robin, and Karen O'Brien. 2008. *Environmental Change and Globalisation: Double Exposure.* Oxford: Oxford University Press.

Lewis, James. 1999. *Development in Disaster Prone Places: Studies of Vulnerability.* London: Intermediate Technology Publications.

Lomborg, Bjørn. 2001. *The Skeptical Environmentalist: Measuring the Real State of the World.* Cambridge: Cambridge University Press.

Lovelock, James. 2006. *The Revenge of Gaia: Why the Earth is Fighting Back – and How We Can Still Save Humanity.* London: Allen Lane.

Luna, Emmanuel. 2001. "Disaster Mitigation and Preparedness: The Case of NGOs in the Philippines." *Disasters* 25 no. 3: 216–226.

MacKinnon, Danny, and Kate D. Derickson. 2012. "From Resilience to Resourcefulness: A Critique of Resilience Policy and Activism." *Progress in Human Geography* 37 no. 2: 253–270.

Manyena, Siambabala. 2006. "The Concept of Resilience Revisited." *Disasters* 30 no. 4: 433–450.

Maskrey, Andrew. 1989. *Disaster Mitigation: A Community Based Approach.* Oxford: Oxfam.

McCarthy, Francis X., and Natalie Keegan. 2009. *FEMA's Pre-disaster Mitigation Program: Overview and Issues.* Washington, DC: CRS report for Congress. 7-5700. Congressional Research Service.

Mercer, Jessica. 2010. "Disaster Risk Reduction or Climate Change Adaptation: Are We Reinventing the Wheel?" *Journal of International Development* 22 no. 2: 247–264.

Mitchell, Timothy. 2011. *Carbon Democracy: Political Power in the Age of Oil.* London: Verso.

Mohan, Giles, and John Mohan. 2002. "Placing Social Capital." *Progress in Human Geography* 26 no. 2: 191–210.

Mosse, David. 2001. "'People's Knowledge', Participation and Patronage: Operations and Representations in Rural Development." In *Participation – The New Tyranny?*, edited by Uma Kothari, and Bill Cooke, 16–35. London: Zed Books.

Nakagawa, Yuko, and Rajib Shaw. 2004. "Social Capital: A Missing Link to Disaster Recovery." *International Journal of Mass Emergencies and Disasters* 22 no. 1: 5–34.

North Atlantic Treaty Organization. 2001. *NATO's Role in Disaster Assistance.* Brussels: NATO Civil Emergency Planning, Euro-Atlantic Disaster Response Coordination Centre.

Nickel, Patricia M., and Angela M. Eikenberry. 2007. "Responding to 'Natural' Disasters: The Ethical Implications of the Voluntary State." *Administrative Theory and Praxis* 29 no. 4: 534–545.

O'Keefe, Phil, Ken Westgate, and Ben Wisner. 1976. "Taking the Naturalness Out of Natural Disasters." *Nature* 260: 566–567.

Pelling, Mark. 2003. *The Vulnerability of Cities: Natural Disasters and Social Resilience.* London: Earthscan.

Pelling, Mark. 2011. *Adaptation to Climate Change: From Resilience to Transformation.* London: Routledge.

Pelling, Mark, Karen O'Brien, and David Matyas. 2015. "Adaptation and Transformation." *Climatic Change* 133 no. 1: 113–127.

Reid, Julian. 2012. "The Disastrous and Politically Debased Subject of Resilience." *Development Dialogue* 58: 67–79.

Ribot, Jesse. 2011. "Vulnerability before Adaptation: Towards Transformative Climate Action." *Global Environmental Change* 21 no. 4: 1160–1162.

Ribot, Jesse. 2014. "Cause and Response: Vulnerability in Climate in the Anthropocene." *Journal of Peasant Studies* 41 no. 5: 667–705.

Rocha, José Luis, and Ian Christoplos. 2001. "Disaster Mitigation and Preparedness on the Nicaraguan Post-Mitch Agenda." *Disasters* 25 no 3: 240–250.

Sommer, A., and W. Mosley. 1972. 'East Bengal Cyclone of November, 1970: Epidemiological Approach to Disaster Assessment'. *The Lancet* 13 May: 7–8.

Summers, Lawrence, and Lant Pritchett. 1993. "The Structural Adjustment Debate." *The American Economic Review* 83 no. 2: 383–389.

Szerszynski, Bronislaw, and John Urry. 2010. "Changing Climates: Introduction." *Theory, Culture and Society* 27 no. 2–3: 1–8.

United Nations. 1992. *United Nations Framework Convention on Climate Change.* FCCC/INFORMAL/84. New York, NY: United Nations.

United Nations. 2005. *Hyogo Framework for Action 2005–2015: Building the Resilience of Nations and Communities to Disasters.* Extract from the Final Report of the World Conference on Disaster Reduction. A/CONF.206/6. Geneva: United Nations International Strategy for Disaster Reduction.

United Nations International Strategy for Disaster Reduction. 2015. *Sendai Framework for Disaster Risk Reduction.* UNISDR, Geneva. Available at: http://www.unisdr.org/files/43291_sendaiframeworkfordrren.pdf (accessed May 11 2017).

Veltmeyer, Henry. 2005. "Development and Globalisation as Imperialism." *Canadian Journal of Development Studies* 26 no. 1: 89–106.

Vidal, John. 2013. "Poor Countries Walk out of UN Climate Talks as Compensation Row Rumbles on." *The Guardian*, November 20, 2013. http://www.theguardian.com/global-development/2013/nov/20/climate-talks-walk-out-compensation-un-warsaw.

Walker, Jeremy, and Melinda Cooper. 2011. "Genealogies of Resilience: from Systems Ecology to the Political Economy of Crisis Adaptation." *Security Dialogue* 42 no, 2: 143–160.

Watts, Michael. 1993. "Hunger, Famine and the Space of Vulnerability." *GeoJournal* 30 no. 2: 117–125.

Weymouth, Lally. 2013. "Lally Weymouth: an interview with Australia Prime Minister Tony Abbott." *Washington Post.* October 25, 2013. http://www.washingtonpost.

com/opinions/lally-weymouth-an-interview-with-australia-prime-minister-tony-abbott/2013/10/24/f718e9ea-3cc7-11e3-b6a9-da62c264f40e_story.html.

Williams, Glyn. 2004. "Evaluating Participative Redevelopment: Tyranny, Power and (Re) Politicisation." *Third World Quarterly* 25 no. 3: 557–578.

Wisner, Ben. 1993. "Disaster Vulnerability: Scale, Power and Daily Life." *GeoJournal* 30 no. 2: 127–140.

Wisner, Ben. 2001. "Risk and the Neoliberal State: Why Post-Mitch Lessons Didn't Reduce El Salvador's Earthquake Losses." *Disasters* 25 no. 3: 251–268.

Woolcock, Michael. 2001. "The Place of Social Capital in Understanding Social and Economic Outcomes." *Canadian Journal of Policy Research* 2 no. 1: 11–17.

World Bank. 2001. *World Development Report 2000/2001: Attacking Poverty*. New York, NY: Oxford University Press.

World Bank. 2015. *World Development Report 2015: Mind, Society, and Behaviour.* Washington DC: World Bank.

Yong, Chen. 1988. *The Great Tangshan Earthquake of 1976: An Anatomy of Disaster.* Oxford: Pergamon Press.

3

BETWEEN PRECARITY AND THE SECURITY STATE

A post-vulnerability view

Kenneth Hewitt

Vulnerability retrospective

In the latter half of the 20th century, socially derived vulnerability assumed increasing importance in disaster studies. 'Socially derived' refers to exposure and outcomes that depend upon pre-existing human conditions where disaster occurs. In this era, people's vulnerability came to be viewed as largely *manufactured*, a product of social and cultural contexts or histories. Social and political constructions were seen to shape disaster outcomes, rather than just objective hazards or misfortune (Hewitt 1997, 141–168; Bankoff, Frerks, and Hilhorst 2004). Extreme events and crises have remained a concern, but not as primary causes or guides to action.

This chapter outlines the writer's roughly fifty-year experience of vulnerability approaches. Influential studies shifted the balance of explanation and response towards a growing sense that, rather than being the *cause* of harmful conditions, earthquake or storm, explosion, and fire, mainly reveal *already unsafe* persons, sites, and sectors. A cost-benefit view of disaster risk was also challenged, as emphases moved to the roles of inequities across societies. There was movement away from assumptions that rigidly separate disaster from 'normal life' and a pre-occupation with emergency responses was questioned (Hewitt 1983, 22).

Everyone is in some measure vulnerable. However, disasters mainly impact and reveal people who have few, poor or absent protections, and degraded capacities to respond. Moreover, these predicaments tend to arise less, or not at all, from hazards and impacts or out of the crisis itself. They were seen as anthropogenic; to depend mainly on past human developments and conditions of *everyday life*. Even unprecedented catastrophes like the Halifax Harbour explosion of 1917, the 'Titanic' sinking, Chernobyl, Bhopal, and Fukushima, turn upon relatively

DOI: 10.4324/9781003219453-4

ordinary neglect and carelessness, banal preoccupations at the price of safety (Turner 1978; Perrow 1984; Kletz 1993).

Meanwhile, attention to the human roots of disasters brought a growing scepticism about impersonal controls, against Malthusian notions of 'population', and remnants of Social Darwinism and geographical determinism. Equally, appeals to 'Acts of God' or 'Mother Nature' were being questioned. By the 1980s the vulnerability focus was strengthened by an upsurge of studies exploring risk terrains associated with gender, ethnicity, race, and class. Others concerned urban slums and neglected rural people: the predicaments of excluded, colonised and marginalised groups (Watts 1983; Maskrey 1989; Blaikie et al. 1994; Lavell 1994; Kobayashi 1994; Enarson and Morrow 1998; Fernández 1999; Fordham 2012; Katz 2013).

Among geographers, the case for such understanding was prepared by various notions, notably 'choice of use or adjustment to hazards' (White 1961); 'living with risk' (Burton et al. 1978), and 'the social fabric of risk' (Cutter 1993, 21). 'Social behavior' became a focus of DRR for sociologists (Quarantelli 1998), and tended to support Kreps' (1984, 309) radical conclusion that: '...disasters both reveal elemental processes of the social order and are explained by them'. In all, rather than extreme, unstoppable forces, most disaster risk and losses originated in what Arundhati Roy (2019, 16) characterises as: '...*the endless crisis of normality*... policies and processes that make *ordinary things* – food, water, shelter, and dignity – a distant dream for ordinary folks (*emphasis added*)'. Equally, in many cases disaster survivors gave support to Wildavsky's (1981, 19) maxim: 'richer is not merely better, but safer, ... The lower the income, the higher the death rate at every age'.

A preventive view

Related and complimentary observations supported the pursuit of preventive measures. Of course, there are forces or conditions that exceed all reasonable human capacities to protect against. They are, however, few where humans have established permanent settlement. Also, they are rarely decisive in events that make the disasters lists.

In studies of earthquake sites in the Eastern Mediterranean, I became convinced that most damages could be avoided, mitigated, or prevented (Hewitt 1997, 197–224). This was readily apparent from walking the ground in the same areas before, and soon after, seismic events. Impacts could indeed seem terrible and arbitrary, but specifics revealed obviously unsafe sites, neglect of the built environment, careless practices, and failure to obey building codes (Ozerdem and Jacoby 2006). Rather, reports from the same sites are of repeated failures to protect, of a lack of readiness, and of timely warnings or evacuation. The casualties and damage encountered could have been reduced or stopped and by well-known, affordable means – had that been attempted (Davis 1981). Injury and destruction arise mainly through avoidable social flaws; by neglect, accidents, unsafe practices, and lack of resources and services, known and that could be provided. The same conclusions are revealed in dozens of official inquiries and intensive investigations of disaster events (Hewitt 2013, Tables 3 and 4).

It is important to stress the distinction between vulnerability and *exposure* to dangers. Vulnerability is made clear when exposure to dangerous forces occurs. However, people who are more exposed can, in fact, be less vulnerable if provided with adequate protections. Limited exposure may offset unusual weaknesses – what might be termed the trade-offs of frontline responders, firefighters, or lighthouse keepers. Then again, those whose losses stand out in disasters tend to exhibit both adverse exposure and multiple vulnerabilities. In the 2020 CORONA-19 pandemic, for example, social inequities, indifference, or neglect have been the scandalous partners of plague. Greater concentrations of infections and mortality affect persons and groups already disadvantaged. Frontline workers were widely denied adequate protections and support – though called 'essential'. Persons with pre-existing health-related weaknesses and insecure livelihoods have proved more at risk. They include, especially, women, racialised groups, migrant workers, needy children, and the uncared-for elderly.

Such inequities and adverse outcomes have been reported from the USA and India, Russia and Western Europe, Canada, and Brazil despite their many differences. In each case, survival is critically dependant on the quality, or otherwise, of pre-existing health care. Many, if not most, deaths and illness could have been avoided or prevented with established methods. Large differences in outcomes were shown to relate to government policies, for-profit versus public assistance, variable diligence in public affairs. Not the least issues were poorly managed and inadequate policies – failure to follow 'the science', but also confused messaging.

None of this is a surprise for those familiar with the vulnerability perspective on DRR. It reinforces the argument against separating calamity from the rest of life, and, in particular, from the political economy of disadvantage. Vulnerability has increasingly drawn attention to the role of political power (Hannigan 2012). Results point to excluded voices, marginalised communities, and indebted groups, and to conditions embedded in cultural values and practices (Hewitt 1983; Mileti 1999; Krüger et al. 2015; Sudmeier-Rieux et al. 2017). As such, many conclude the focus of DRR needs to be on precautionary and preventive measures, again for those most at risk as identified from the losses in recent disasters. Though still in need of greater adherence, a preventive view has received strong support in some quarters, as expounded in the Hyogo and Sendai Accords (UNDRR 'PreventionWeb', 2020; UNDP 2004; UNISDR 2005, 2010, 2020; Lewis, 2014; Pigeon and Rebotier 2016).

A common complaint is that vulnerability perspectives emphasise human limits and weakness, while ignoring agency. The point is valid, but hardly true of the literature cited here. It includes many accounts of extraordinary responses, both by disaster survivors and resilient communities (Oliver-Smith 1992; Enarson and Morrow 1998; Lewis 1990; Oliver-Smith and Hoffman 2001). Many of those harmed in disasters also do much of the life-saving, sheltering, feeding, evacuation, protective work and, especially, 'loss bearing' (Steinberg 2000; Dekens 2007; Sliwinski 2018; Remes 2016). Meanwhile, if resilience stresses planned actions, vulnerability explores the interwoven physical, ecological, and cultural preconditions of disaster. It emerges uniquely from lives as lived and addresses the systemic and deliberate undermining of peoples' security (Erikson 1976; Hartmann and Boyce 1983; Watts 1983; Hewitt 2007; Klein 2007; Katz 2013).

Nevertheless, a growing preference for other styles of DRR cannot be denied. By 2019 a vulnerability approach seemed to be losing appeal (Adger 2006). The remainder of this chapter addresses what seem post-vulnerability predicaments. They include prevailing views that neglect certain elements of DRR, even as they become more influential elsewhere. First, the importance is examined of a work/labour/employment nexus – *and* lack of attention to it. This is seen as related to precarious living generally. Second, the significance of armed violence is revisited and the vulnerability it can reveal. In addition to 'disasters of war', militarised emergency measures are considered, and the ballooning 'security state'. Finally, the anthropogenic and violent origins of disaster that underscore humanitarian interpretations are interrogated. They challenge the requirements – and lack of – non-violent and peace-building approaches to DRR.

Worlds of work

A job, or lack of one, can be decisive in exposure, impacts and responses to disaster. Unduly affected are those with low income and high unemployment, or in perennially dangerous working conditions. Recent disasters yield many stories of job loss, dangerous work, inability to meet mortgage and other payments. Unemployment and insufficient capacities are tied to two other great drivers of social risk, debt, and usury (Strolovitch 2021). Confirmation comes from many places impacted by the Indian Ocean tsunami (2004), from the financial crisis in North America in 2008, New Orleans after Hurricane Katrina (2005), and the earthquakes in Kashmir (2005 and 2019), Haiti (2010), and Nepal (2015). In New Orleans, racialised, unemployed survivors are prevented from returning to their places of work (Adams 2013). Former jobs were lost to cheaper labour from outside.

Perrow (1984), arguing for an organisation-based understanding of disasters, emphasises employment as the decisive link between most persons and the institutions which attempt to control, and largely do shape safety in the modern world. Employment conditions systematically separate those who are secure and entitled from the vulnerable, and unentitled. Late modernity has brought massive redistribution of risks and benefits, albeit poorly monitored or understood, and cynically accepted (Ericson and Doyle 2003; Briken and Eick 2011).

Around the world are millions rated as 'unemployed' or 'illegal' who nevertheless work hard every waking hour; not least the women whose work, in so many societies, is not recognised as such. In Haiti, 2010, when the earthquake struck, there were the 70 per cent supposedly unemployed. According to Katz (2013), they were constantly engaged in multiple tasks to make a living. The problem requires separation of 'work', from a job that provides a reliable, living wage. Then there is uprooting and break up of communities in disasters, that arise from loss of, but also desperate searches for, employment. Crowds of migrant or trafficked workers emerge in and from disaster zones. Survivors may be abandoned in chronically jobless places, in urban slums, or refugee camps (Bales 2004; Svistova and Pyles 2018).

High levels of workplace deaths, illness, and injury are indicators of unsafe living. The International Labour Organization (ILO 2011) estimates 2.3 million men and women are killed annually by work-related accidents or diseases worldwide. Hazardous substances alone are estimated to cause 651,279 deaths a year. Concentrated disaster losses repeatedly signal areas of labour exploitation: sweat shops and slaughterhouses, fruit and vegetable pickers, garbage collection, construction sites, mining and forestry. Work-related accidents and disease outbreaks especially affect the younger and older cohorts of workers, female and racialised labour, while estimates are flawed by chronic under-reporting (Anderson 2000; Kletz 1993). Conversely, where monitored and competently assessed, most workplace casualties are also deemed 'predictable and preventable' (Turner 1978; Perrow 1984; Pigeon and Rebotier 2016).

On one level, these issues are pervasive, or appear self-evident. Yet, most studies ofDRR show little interest in following up on work-related dangers, employment or lack of it. I suggest this hinders a post-vulnerability discussion. In other fields, labour is recognised as the epicentre of (in)security. Of late, in the context of globalisation, the topic has become engaged with fast-changing and precarious living.

Precarity

This notion has its problems and critics (Standing 2011; Bales 2004; Butler 2004; Waite 2009) Yet, its absence from most disaster studies is puzzling, and compounds a mainstream failure to engage with other notions that help expose modern dangers and material uncertainty. They include 'the risk society' of Beck (1992), 'the accident' of Virilio (2007), and Perrow's (1984) 'normal accidents.' Each is relevant for a post-vulnerability characterising late modern disasters. Add *precarity* to the list.

Early thought on the topic arose from religious and social movements concerned with poverty and destitution. Recent debates emphasise neoliberal overturning of work and labour rights, negotiated in wealthier industrial countries after World War II. 'Vulnerable employment' is a major issue, defined as: '... precarious work that places people at risk of continuing poverty and injustice resulting from an imbalance of power in the employer-worker relationship' (TUC 2008). However, while labour is central, others believe that: '...precarity is not just about work. It is about the precaritisation of life along many different fields;

housing, women's rights, education, health care, social rights, culture, mobility and migration' (Trimikliniotis 2017, 227). Bourdieu (1963) adopted the term *précarité* for employment in the Algerian context, and to identify temporary or casual workers. They endure the worst of a racialised, colonial division of labour; that is, in contrast to a permanently employed Francophone, 'white' work force. The fundamentals of such precarity were ever-present across former colonies, and largely persist in so-called 'independent' or post-colonial settings.

Precarious lives have had the most massive growth under globalisation, through strategies to exploit cheap labour. It also continues to adopt and reconfigure pre-existing systems of caste, feudal exploitation, and the enslavement of workers. A related theme, 'labour *arbitrage*', is relevant here. It includes out-sourcing or moving physical plant or migrants to countries with low pay, and where worker and environmental protections are few. And the same marks populations beset by multiplying dangers and losses associated with social upheavals. Precarity is widely associated with rapid urbanisation and megacities. This is most evident in what Davis (2006) calls the 'planet of slums', whose problems overlap with all other forms of disaster risk and the worst of unemployment contexts. These involve insecure residence, organised crime and trafficking, child labour, habitat damage, repression, and armed conflict.

Such conditions and precarious lives recur across reports of disaster and in studies of DRR (Maskrey 1989; Bohle 1993; Lavell 1994; Hewitt 2000, 2013; Schuilenburg 2012; Hilhorst 2013; Lewis 2014). Employment debates and DRR must consider the massive growth of unprotected labour. These precarious lives are drawn to hazardous locations in and near places of work. In disasters at Bhopal (1984), Chernobyl (1986), Fukushima (2011), the Grenfell Tower (2017), and the explosion at Beirut (2020) work-related or unemployment stresses amplified destruction, lives lost, and livelihoods.

Equally relevant is the question of how people who avoid or recover readily from disasters do so. It may be thanks mainly to established assets: income or credit, welfare or benefit provisions, that in turn are usually obtained with or tied to employment. Beyond so-called underlying risk factors, these highlight the existence and roles of 'underlying *safety factors'*. They show how security can also turn upon inherited or class benefits, churches or temples, Friendly or Benevolent Societies. These mediate between work, welfare, and disaster risk. They also demand greater attention to people's pre-existing security, or degrees of alienation. This arena can also reveal mutualism and safety cultures that help protect the more vulnerable – and underscore its absence elsewhere. In fact, precarity discourse has also become engaged with worker flexibility, resistance, and collective action. The debate comes laden with stories of resistance and cooperation among workforces. There are movements to assist and empower women and people of colour, migrants and seasonal labour, benefits that can spill over into DRR. As with getting and holding a job, established social supports can be more important than skills or readiness, even wealth *per se*.

Discussions of precarity suggest a major fork in the road for DRR, relating to social risk and changes in work environments and practices. Multiple and sharp differences in material security have been uncovered. Preventive alternatives promote, or should promote, a commitment to welfare provisions tied to civil responsibilities. There are the special forms and degrees of precarity of frontline workers, both in and beyond disaster zones. Vulnerability is found to turn on needs such as sick leave with pay, disability provisions, a universal minimum income, and pay equity for women. It may involve access to housing, daycare, schooling, and property insurance. There is a conspicuous absence of these concerns, even in responses of DRR.

Armed violence

Historically, war and rebellion were placed among the four great calamities and seen as integral to understanding disaster (Sorokin 1942). Recently, however, they have been largely by-passed in DRR. In general, issues of armed violence are absent from the Sendai Accords, and most of the disaster community literature. Wisner (2012) provides a compelling overview of how violent conflict affects *other* types of disaster, mainly events assigned to environmental hazards and disease. There are studies that recognise the importance of armed force in Complex Humanitarian Emergencies (Rangasami 1987; Edkins 2000). Some other research reinforces the case being made here (Loyd et al. 2012; Hilhorst 2013). They are exceptions, too, that help reveal how armed force and the resources devoted to it remain a major source of calamity and of the singular risks identified with the Anthropocene.

For me, geographies of war were critical in establishing the scope and significance of social vulnerability, based on the investigation of air raids against cities in World War II, and comparing Britain, Germany, and Japan. These were the three most heavily bombed countries, and among the few major air powers carrying out great raids. Their combined civilian deaths exceeded 1.5 million. Severe injuries affected 2.3 million. Over 17 million city dwellers lost their homes (Hewitt 1997, 216–319). It was an epoch bracketed by the 1915 Zeppelin fire raids on London, the attack on Guernica during the Spanish Civil War, and involved the carpet bombing of Coventry, Dresden, Nagasaki, and the unmanned V-rocket raids (Lindqvist 2001; Funck and Chickering 2004). The larger raids far exceeded losses in most events classed as 'natural' or 'technological' disasters. Twelve raids on German cities each exceeded 3,000 civilian deaths, with a maximum of 40,000 at Hamburg between July 27 and 28 in 1943. The Tokyo fire raid of March 9, 1945, with over 100,000 civilian deaths, was the most lethal, its casualties exceeding even Hiroshima (Hewitt 1997; Primoratz 2010).

A particularly horrendous case was the September 9, 1944, firestorm attack on the German city of Darmstadt (Schmidt 1964). Out of 10,550 deaths, 58 per cent were female and 33 per cent children. The raid created a firestorm and accounted for 95 per cent of wartime raid deaths in the city. As an old cultural centre with many medieval, inflammable buildings, it burned rapidly. Its raid shelters gave rise to some of the largest proportions of 'oil age' casualties in the fire raids; people who died of heatstroke, carbon monoxide poisoning, and incineration. Fire raids were the most destructive in all three countries (Hewitt 1993; Friedrich 2006). From the beginning and in each country, the attacks were called 'terror raids' (Hewitt 1994a). They introduced a dimension of disaster that would be called 'urbicide' (Graham 2004, 165). They revealed the defencelessness of cities against air power, and the return of urban mass fires as a uniquely wartime threat (Bankoff 2012).

Evidence and personal testimonies from civilian survivors of the Second World War bombing greatly influenced my sense of vulnerability in disaster (Hewitt 1993, 1994a). Post-war investigations detailed the civilians' plight, and the atrocities largely denied in wartime propaganda (USSBS E 1945a, 1945b, 1947b; USSBS P 1947a, 1947b; Bidinian 1976; Harrisson 1976). What these sources provided and depended on disclosing was a 'view from below', experience on the ground and beneath the bombs. Unlike the more familiar war histories of men in uniform and the fighting fronts, the story here comes almost entirely from civilians, with women commonly the main witnesses, participants, and victims. In another divergence from the soldiers' wars, city dwellers are encountered together, and not only as non-combatants but as citizen and resident groups, relatives, and neighbours, old and young, families and communities. They also comprised the frontline workers of civil life – nurses, bakers of bread, teachers, and clerks.

Disproportionate casualties identified the most vulnerable city folk – children and the elderly, prisoners and the disabled. More women died than men in the worst raids on Britain and, in total, in Germany and Japan (Hewitt 1993, 32–37). Ten times as many women were killed as airmen sent against them. The single largest class of raid victim was registered as 'housewife'. A smaller but tragic share were raid evacuees, mainly children, young women, and the elderly, transferred to supposedly safe havens and undefended cities, many of which were attacked and proved lethal. There could be also crowds of POWs and forced labour cadres. As at Dresden, refugees from previous raids were encamped or hiding in and around the bombed cities. There were clusters of persons, classes, and sectors, featured among the homeless, displaced, and otherwise deprived survivors. It reveals how the profiles of those at risk differed from usual war zones but were not dissimilar from other disasters.

The scale of destruction and the death tolls reflect not only the incendiary and high explosive weapons used but also the pre-existing urban layout and social differentiation. The most calamitous attacks were those that achieved massive concentrations of bombs in congested, inflammable, and densely populated urban areas. In all three countries, the worst attacks were called 'slum raids'. They targeted inner-city neighbourhoods and congested areas where the industrial working class lived (Hewitt 1994b). Many victims were families of soldiers at the front.

Domestic life was most at risk and harmed. Experience under the bombs underscored forms of vulnerability critical for civilians and caused exceptional damage to civil life support (Hewitt 1994a, 2013). Various developments specially threatened civil and domestic life. For example, an air force preference for night raids, helped minimise the loss of planes and aircrews but brought more indiscriminate, ill-aimed attacks that posed severe problems for civilians on the ground.

Air raid shelters, in many cases, were reported as noisy and insanitary; damp, cold or stuffy places where contagious diseases were spread (USSBS E 1945b; USSBS P 1947b; Harrisson 1976; Hewitt 1994a). Those needing shelter were kept awake by frightened, screaming children or rude and loud neighbours. When asked about their raid experience, bombing survivors in most cities complained of weariness, loss of sleep, and skin diseases. They spoke of hunger, poorly prepared food, and the total disruption of everyday life. One sees how these stresses undermined their abilities to respond and stay alert. A predominance of 'false alarms' – sirens that woke everyone up but were not followed by an attack – led many people to stay in their home. There, however, when a major raid did hit the city, some of the worst death tolls were recorded, not least, among concentrations of persons with the same or related surnames.

After 1945, air attacks continued bringing mass death and devastation almost worldwide (Ahlström and Nordquist 1991; Herold 2003). They have added to the list of urbicides; Pyongyang, Hanoi, Aleppo, Grozny, Halabjain, Falluja, Hargeisa, Herat, Raqqa, Sa'na, – and Sarajevo where the term was coined (Graham 2004, 138). In the drylands of the Middle East, high explosive bombs have again dominated the civilian experience, rather than incendiaries as weapons of choice.

A further reason to revisit air war involves the *civil defence* origins of disaster management, especially from the 'home fronts' of the Second World War (O'Brien 1955; Harrisson 1976). The 'occasion instant' of Mack and Baker (1961) characterised an early preoccupation in disaster management with forces for rapid deployment and emergency measures, and during the Cold War, civil defence remained the official approach to disaster management (NATO-OTAN 2001, 5). The 9/11 attacks provided further impetus, and to a militarised response (Hilhorst 2013).

Civil defence

Homeland Security is essentially a much-enlarged civil defence system, one integral to today's security state (Hewitt 2017; Nguyen 2016). With little discussion, emergency readiness has been merged with a whole array of security-related agencies. It is positioned in terms of conflict priorities, especially the war on terror (Schuilenburg 2012). Draconian policing seeks to control civil protests, travellers, migrants and their labour, and asylum seekers. Official disaster management recalls President Eisenhauer's 1961 'Military Industrial Complex', some identifying a 'Disaster Industrial Complex' (Svistova and Pyles 2018) or a 'Security Industrial Complex' (Macdonald 2006). Once more, disaster management finds its home (-land) amid violent threats, and a 'logic-of-war' (Gilbert 1998, 12–13; Hilhorst 2013, 5).

However, the most fateful lesson of civil defence in the world wars has been ignored: its general inadequacies and failure when most needed. Despite many instances of bravery and self-sacrifice, in the great raids from Coventry to Dresden and Tokyo, civil defence forces and facilities were totally overwhelmed (USSBS E 1945a, 1947a; USSBS P 1947a, 1947b; O'Brien 1955). Air attacks became prescriptions for annihilation. Similarly, the message from most air wars since 1945 is their failure to protect civilians. In the present time of megacities and majorities of urbanised humanity, no cities are remotely provided with means to defend or protect themselves, least of all against the air weapons and strategies being prepared for use against them. Civilians still have the greatest share of violent deaths where raids occur, of disablement, forced uprooting, and refugee struggles (Loyd, 2012). Also, military forces and defence systems assume ever-greater roles in public safety, including in monitoring and disaster response. In adopting a civil defence posture, responsible agencies turn away from the pre-disaster social causes of insecurity (Schuilenburg 2012).

Here, armed force appears like 'the elephant in the room' of public safety and DRR. Vastly more wealth has been poured into air forces than civil protections, into militarised security than public safety. Worldwide, some 30 million people serve in the armed forces; over 75 million if paramilitary and reserves are counted. Police forces, ever more likely to be heavily armed, exceed 12 million. Global military spending reached $1,917 billion in 2019 (SIPRI 2021). That is far above expenditures on DRR. Billions of dollars go for secret electronic surveillance and other intelligence services, into policing, crowd control, carceral systems, and border walls. Military forces assume ever-greater roles in disaster responses. Issues like precarity, even as they play a larger role, are pushed to the back of the queue. It encourages a return to hazards and emergency measures approaches, and away from vulnerability and preventive approaches.

Reconstruction after disaster

Today's repeated calls for 'building back better' raise issues of the air war and DRR in relation to urban reconstruction (Edgington 2011). In the three countries introduced above, the scale of reconstruction far exceeded any others, once seriously undertaken. Within two decades one hardly noticed reminders of war damage. Most of the hectares of rubble were gone, removed, recycled, or built over. In the process, city-folk, planners, and construction companies struggled with the problems and opportunities of urban re-planning and renewal. This also speaks to one of several themes of the so-called disaster cycle. If civil defence covered 'readiness and response', another large element was post-disaster recovery. But the vast subject of 'reconstruction after war' has been largely neglected, as has the huge trove of experience derived from it.

One development seems anachronistic at best, cynical at least, but also echoes what is found happening in disaster zones. Hitherto, I have treated the air raids as unmitigated calamities, atrocities even. I see no way to ignore the mass death, broken bodies, ruins, fires, and uprooting, annihilation of popular sites and heritage. But some very influential people expressed and pursued an entirely different vision. They saw the wartime razing and emptying of the cities as a gift, a 'godsend' or 'blessing in disguise' (Düwel and Gutschow 2013). Mostly these were the voices of politicians and planners, engineering architects, journalists, and public intellectuals of a certain persuasion. For them, the cities destroyed had been living hells, a whole urban world seen only through the lens of slums and decadence; seen as decayed and ruined civilisations (Mumford 1947; Nipper, Nutz and Wiktorin 1993).

Three radically different and contradictory views confront a critical analysis here. First, there were the airmen, who rated the raids mentioned here as military '*successes*' – the greater the death toll and destruction of the city, the more so. Conversely, the raid talk of civilians on the ground, was of terror and grief, of calamities, end-of-the-world feelings, irreversible losses. Finally, as just encountered, there were urban specialists for whom the levelling of the old city presented a golden opportunity. Great improvements were expected by starting from scratch, As the survivors of Coventry and east London mourned their dead, some leading figures just saw huge opportunities to plan and begin to build these and others as the 'city of the future'.

It was recorded how several million civilians died during World War II (Hewitt 1997). Tens of millions more lost their homes and livelihoods, along with much of the finest architecture in the land, now in ruins. But town planners saw a chance to rebuild the city without the nagging constraints of buildings from the past, inhabitants in place, and patterns of ownership; a process that began, as Düwel and Gutschow (2013) recount in stunning detail, while the war and bombing were still going on.

This is a story whose full message of reconstruction has, in many ways, been lost in the post-Hiroshima geostrategic world. Now, the absolute vulnerability of any form of urban building keeps overtaking the latest means to attack them. As an invitation to argue about post-disaster reconstruction after World War II, however, air raid survivors had no lack of arguments to see the raids as anything but 'a good thing'. For some, even the champions of airpower, the raids did not really change the course of the war. They never paid for the costs and deaths in their execution. The resources demanded by the bomber fleets were immense. City-dwellers proved more resilient than air plans expected, less easily cowed. Many more survived than were lost. Finally, others have pointed persuasively to the ugliness and dysfunctional aspects of so much post-war rebuilding. Some say that few periods of architecture and city-planning have been as sterile, authoritarian, and unimaginative as the decades after World War II (Hewitt 1994a).

More pointedly, civilian losses, in air attacks during World War II and subsequently record an almost universal failure to enforce 'civilian immunity' in air war (Gutman et al. 2007, 103–107). Belligerents continue to ignore the rights of 'protected persons.' And this is to make the case for an essentially moral dimension to these disasters, and what civil recovery from them requires. Ethical values and protections have been raised whether and where the more powerful states insist upon enforcement of humanitarian principles and rules – or, most often, do not. Despite the Nuremberg and Tokyo trials, the Nuremberg Code, and the ICC, such violations have disfigured every war since 1945. In this context, the urban attacks of the World Wars are not so much the worst disasters, as among the gravest crimes of war (D'Angostino 1991; Garrett 1993). Again, the raids are disasters that can only be mitigated or prevented through adherence to International Humanitarian Law (Gutman et al. 2007, 22).

Arguably, both raid plans and reconstruction afterwards involved moral and legal choices, the unique and contradictory implications of acts of violence and survival. The calamities are not primarily physical; nor are they only violations of the 'peace', but violations of the legitimate values and justifications of warfare itself. Instead, as it turns out, bombing civilians joins a whole new vocabulary of atrocities that signal the criminality of such actions – domicide, ecocide, gendercide, urbicide, or genocide. Each has brought disasters to defenceless people. Notwithstanding attempts to punish the huge crimes against civilians in the World Wars, most of the principles or 'laws of war' have been violated in every war since 1945 (Gutman et al. 2007). The moral dimensions are also aggravated by widespread impunity claims. Meanwhile, efforts to protect and defend non-combatants highlight the man-made nature of these threats, and show them to be, in principle, preventable by choice.

A further way to address the disasters of war in relations to other hazards could be developed from Gilbert White's well-known commitment to non-violence, an approach that includes central roles for humanitarian law, conscientious objection and peace-making. It would require the banning of indiscriminate attacks and dubious weapons. Civil protections would come from initiatives in public safety, from precautionary, mutualist, and peacekeeping efforts. A rare sign of such an interest emerges with 'disaster diplomacy' (Kelman 2012). True, so many crises, conflicts, and complex emergencies do lack interest in, or are caused by failures of diplomacy. That should change. For the most part, however, these are matters absent from and ignored in the present DRR.

Concluding remarks

My work in the past few decades highlights challenges for DRR that arose from recognition of the widespread and socially manufactured nature of vulnerability and absent protections. In the same time frame, studies of vulnerability shifted the balance of explanation and concern from natural and technological hazards to anthropogenic factors, especially the preconditions of social insecurity and disadvantage. As such, the exit from vulnerability perspectives now taking place needs to be thought through, recognising the moral implications. To address these disasters and human inputs is especially about ethics. Emphasis has moved to the predicaments of people exposed to dangers through work, housing and land uses who face violent repression but lack protections or a voice in public affairs. Environmental abuses, including climate change, are shown to come down, ultimately, to human choices. They arise, fundamentally, from human decisions, values, and rights.

I suspect a major factor that the exit from vulnerability perspectives will encourage is a narrowing of the scope and responsibilities of DRR. Along with placing managerial and administrative tasks first, there has been a moving away from many, once central disaster issues: notably investigations and contexts of dearth and famine, disease and social prejudice, large accidents, and armed violence. This may well come from 'turf wars' or accommodations in government and international agencies that need more critical debate. It coincides with 'push back' against an emphasis on the increasingly anthropogenic sources of modern disasters. In all, the late modern disaster risk environment sits uneasily between the threats of precarious living and the machinations of the security state, including efforts to exclude moral dimensions from DRR.

References

Adams, Vincanne. 2013. *Markets of Sorrow, Labors of Faith: New Orleans in the Wake of Katrina*. Durham, SC: Duke University Press.

Adger, W. Neil. 2006. "Vulnerability." *Global Environmental Change*, 16 no. 3: 268–281. doi:10.1016/j.gloenvcha.2006.02.006.

Ahlström, C. and K.-A. Nordquist. 1991. *Casualties of Conflict: Report for the World Campaign for the Protection of Victims of War*. Uppsala: Department of Peace and Conflict Research, Uppsala University.

Anderson, Bridget J. 2000. *Doing the Dirty Work? The Global Politics of Domestic Labour*. London: Zed Books.

Bales, Kevin. 2004. *Disposable People: New Slavery in the Global Economy*. Berkeley: University of California Press.

Bankoff, Greg, Georg Frerks, and Dorothea Hilhorst, eds. 2004. *Mapping Vulnerability: Disasters, Development, and People*. London: Earthscan.

Bankoff, Greg, Uwe Lübken, and Jordan Sand, eds. 2012. *Flammable Cities: Urban Conflagration and the Making of the Modern World*. Madison: University of Wisconsin Press.

Beck, Ulrich. 1992. *Risk Society: Towards a New Modernity*. Translated by Mark Ritter. London: Sage Publications.

Bidinian, Larry J. 1976. *The Combined Allied Bombing Offensive Against the German Civilian: 1942–1945*. Lawrence, Kansas: Coronado Press.

Blaikie, Piers, Terry Cannon, Ian Davis, and Ben Wisner. 1994. *At Risk: Natural Hazards, People's Vulnerability and Disasters*. London: Routledge.

Bohle, Hans-Georg, ed. 1993. *Worlds of Pain and Hunger: Geographical Perspectives on Disaster Vulnerability and Food Security*. Freiburg Studies in Development Geography. Saarbrücken: Verlag Breitenbach.

Bourdieu, Pierre. 1963. *Travail et Travailleurs en Algérie*. Recherches méditerranéennes. Documents 1. Paris: Mouton.

Briken, Kendra, and Volker Eick. 2011. "Policing the Crisis–Policing in Crisis." *Social Justice: A Journal of Crime, Conflict and World Order*, special issue, 38 (1–2): 1–12.

Burton, Ian, Robert W. Kates, and Gilbert F. White. 1978. *The Environment as Hazard*. New York: Oxford University Press.

Butler, Judith. 2004. *Precarious Life: The Powers of Mourning and Violence*. London: Verso.

Cutter, Susan L. 1993. *Living with Risk: The Geography of Technological Hazards*. London: Edward Arnold.

D'Angostino, Brian. 1991. Review of *The Genocidal Mentality: Nazi Holocaust and Nuclear Threat*, by Robert J. Lifton and Eric Markusen. *Political Psychology* 12, no. 2: 383–385. doi:10.2307/3791472.

Davis, Ian, ed. 1981. *Disasters and the Small Dwelling*. Oxford: Pergamon Press.

Davis, Mike. 2006. *Planet of Slums*. New York: Verso.

Dekens, Julie. 2007. *Local Knowledge for Disaster Preparedness: A Literature Review*. Kathmandu: International Centre for Integrated Mountain Development.

Düwel, Jörn, and Nils Gutschow. 2013. *A Blessing in Disguise: War and Town Planning in Europe 1940–1945*. Berlin, DOM Publishers.

Edgington, David W. 2011. "Viewpoint: Reconstruction after Natural Disasters: The Opportunities and Constraints Facing our Cities." *Town Planning Review* 82, no. 6: v–xi. https://www.jstor.org/stable/41300364

Edkins, Jenny. 2000. *Whose Hunger? Concepts of Famine, Practices of Aid*. Borderlines 17. Minneapolis: University of Minnesota Press.

Enarson, Elaine Pitt, and Betty Hearn Morrow. 1998. *The Gendered Terrain of Disaster: Through Women's Eyes*. Westport, CT: Praeger.

Ericson, Richard V., and Aaron Doyle, eds. 2003. *Risk and Morality*. Toronto: University of Toronto Press.

Erikson, Kai. 1976. *Everything in its Path: Destruction of Community in the Buffalo Creek Flood*. New York: Simon and Schuster.

Fernández, Manuela, ed. 1999. *Cities at Risk: Environmental Degradation, Urban Risks and Disaster in Latin America*. The Network for Social Studies on Disaster. Lima: LA RED.

Fordham, Maureen. 2012. "Gender, Sexuality and Disaster." In *The Routledge Handbook of Hazards and Disaster Risk Reduction*, edited by Ben Wisner, J.C. Gaillard, and Ilan Kelman, 424–435. London: Routledge.

Friedrich, Jörg. 2006. *The Fire: The Bombing of Germany1940–1945*. Translated by Allison Brown. New York: Columbia University Press.

Funck, Marcus and Roger Chickering, eds. 2004. *Endangered Cities: Military Power and Urban Societies in the Era of the World Wars*. Boston: Brill Academic.

Garrett, Stephen A. 1993. *Ethics and Air Power in World War II: the British Bombing of German Cities*. New York: St. Martin's Press.

Gilbert, Claud. 1998. "Studying Disasters: Changes in the Main Conceptual Tools". In *What is Disaster? A Dozen Perspectives on the Question*, edited by E.L. Quarantelli, 11–18. London: Routledge.

Graham, Stephen, ed. 2004. *Cities, War, and Terrorism: Towards an Urban Geopolitics*. Oxford: Wiley-Blackwell.

Gutman, Roy, David Reiff, and Antony Gary Dworkin. 2007. *Crimes of War: What the Public Should Know 2.0*. Revised edition. New York: W.W.Norton.

Hannigan, John. 2012. *Disasters Without Borders: The International Politics of Natural Disasters*. London: Polity.

Harrisson, Tom. 1976. *Living Through the Blitz*. London: Collins.

Hartmann, Betsy, and James K. Boyce. 1983. *A Quiet Violence: View from a Bangladesh Village*. London: Zed Press.

Herold, Marc W. 2003. *Blown Away; Myth and Reality of Precision Bombing in Afghanistan*. Monroe, ME: Common Courage Press.

Hewitt, Kenneth. 1983. "The Idea of Calamity in a Technocratic Age." In *Interpretations of Calamity from the Viewpoint of Human Ecology*, edited by Kenneth Hewitt, 3–32. London: Allen and Unwin.

Hewitt, Kenneth. 1993. "Reign of Fire: The Civilian Experience and Urban Consequences of the Destruction of German Cities, 1942–1945." In *Kriegzerstörung und Wiederaufbau deutscher Städte*, edited by Joseph Nipper and Manfred Nutz. *Kölner Geographische Arbeiten* 57: 25–45. Köln: Geographisches Institut der Universität zu Köln.

Hewitt, Kenneth. 1994a. "When the Great Planes Came and Made Ashes of our City...: Towards an Oral Geography of the Disasters of War." *Antipode* 26, no. 1: 1–34.

Hewitt, Kenneth. 1994b. "Civil and Inner City Disasters: The Urban Social Space of Bomb Destruction" *Erdkunde* 48, no. 4: 259–274.

Hewitt, Kenneth. 1997. *Regions of Risk: A Geographical Introduction to Disasters*. Edinburg Gate, Harlow: Addison Wesley Longman.

Hewitt, Kenneth. 2000. "Safe Place or 'Catastrophic Society'? An Overview of Risk and Disasters in Canada." *Canadian Geographer Special Issue*, "Risk and Hazards in Canada," 44, no. 4: 325–341.

Hewitt, Kenneth. 2007. "Preventable Disasters: Addressing Social Vulnerability, Institutional Risk and Civil Ethics." *Geographische Rundschau International Edition* 3, no. 1: 43–52.

Hewitt, Kenneth. 2013. "Total War Meets Totalitarian Planning: Some Reflections on Königsberg/Kaliningrad." In *A Blessing in Disguise: War and Town Planning in Europe 1940–1945*, edited by Jőrn Düwel and Nils Gutschow, 88–103. Berlin: DOM Publishers.

Hewitt, Kenneth. 2017. "Disaster Risk Reduction (DRR) in the Era of 'Homeland Security': The Struggle for Preventive, Non-violent, and Transformative Approaches." In *Emerging Issues in Disaster Risk Reduction, Migration, Climate Change and Sustainable Development*, edited by Karen Sudmeier-Rieux, Manuela Fernández, Ivanna M. Penna, Michel Jaboyedoff, and J.C. Gaillard, 35–52. Heidelberg: Springer.

Hilhorst, Dorothea, ed. 2013. *Disaster, Conflict and Society in Crisis*. London: Routledge.

ILO. 2011. "Occupational Safety and Health: World Statistics." 1996–2021 International Labour Organization (ILO). Last modified July 13, 2011.. https://www.ilo.org/moscow/areas-of-work/occupational-safety-and-health/WCMS_249278/lang--en/index.htm.

Katz, Jonathan M. 2013. *The Big Truck Went By: How the World Came to Save Haiti and Left Behind a Disaster*. New York: St. Martin's Press.

Kelman, Ilan. 2012. *Disaster Diplomacy: How Disasters Affect Peace and Conflict*. Florence: Routledge.

Klein, Naomi. 2007. *The Shock Doctrine: The Rise of Disaster Capitalism*. Toronto: Alfred A. Knopf.

Kletz, Trevor A. 1993. *Lessons from Disaster: How Organizations Have No Memory and Accidents Recur*. Rugby, Warwickshire: Institution of Chemical Engineers.

Kobayashi, Audrey, ed. 1994. *Women, Work and Place*. Montreal: McGill-Queens University Press.

Kreps, Gary A. 1984. "Sociological Enquiry and Disaster Research." *Annual Review of Sociology* 10: 309–330.

Krüger, Fred, Greg Bankoff, Terry Cannon, Benedikt Orlowski, and E. Lisa F. Schipper. 2015. *Cultures and Disasters: Understanding Cultural Framings in Disaster Risk Reduction*. London: Routledge.

Lavell, Allan, ed. 1994. *Viviendo en Riesgo: Comunidades Vulnerables y Prevencion de Desastres en America Latina*. Bogota: LA RED/FLACSO.

Lewis, James. 1990. "The Vulnerability of Small Island States to Sea Level Rise: The Need for Holistic Strategies." *Disasters* 14 no. 3: 241–249.

Lewis, James. 2014. "The Susceptibility of the Vulnerable: Some Realities Reassessed." *Disaster Prevention and Management* 23, no.1: 2–11.

Lindqvist, Sven. 2001. *A History of Bombing*. New York: The New Press.

Loyd, Jenna, Matt Mitchelson, and Andrew Burridge, eds. 2012. *Beyond Walls and Cages; Prisons, Borders and Global Crisis*. Athens: University of Georgia Press.

Macdonald, Heather. 2006. "Commentary: The Security-Industrial Complex." Manhattan Institute for Policy Research, Inc. https://www.manhattan-institute.org/html/security-industrial-complex-1539.html

Mack, Raymond W., and George W. Baker. 1961. *The Occasion Instant: The Structure of Social Responses to Unanticipated Air Raid Warnings*. Disaster Study 15, Publication 945, Disaster Research Group. Washington DC: National Academy of Sciences-National Research Council.

Maskrey, Andrew. 1989. *Disaster Mitigation: A Community Based Approach*. Oxford: Oxfam.

Mileti, Dennis. 1999. *Disasters by Design: A Reassessment of Natural Hazards in the United States*. Washington, DC: Joseph Henry Press.

Mumford, Lewis. 1947. *City Development: Studies in Disintegration and Renewal*. London: Secker and Warburg.

North Atlantic Treaty Organization. 2001. *NATO's Role in Disaster Assistance*. Brussels: NATO Civil Emergency Planning, Euro-Atlantic Disaster Response Coordination Centre.

Nguyen, Nicole. 2016. *A Curriculum of Fear: Homeland Security in U.S. Public Schools*. Minneapolis: University of Minnesota Press.

Nipper, Joseph, Manfred Nutz, and Dorothea Wiktorin, eds. 1993. *Kriegszerstörung und Wiederaufbau deutscher Staedte. Kölner Geographische Arbeiten* 57. Köln: Geographisches Institut der Universität zu Köln.

O'Brien, Terence. 1955. *Civil Defence*. London: Her Majesty's Stationary Office, Longmans, Green and Co. Ltd.

Oliver-Smith, Antony. 1992. *The Martyred City: Death and Rebirth in the Andes*. 2nd edition. Prospect Heights IL: Waveland Press.

Oliver-Smith, Antony, and Suzanna M. Hoffman, eds. 2001. *The Angry Earth: Disaster in Anthropological Perspective*. 2nd edition. New York: Routledge.

Ozerdem, Alpasian, and Tim Jacoby. 2006. *Disaster Management and Civil Society: Earthquake Relief in Japan, Turkey and India*. London: I.B. Tuaris.

Perrow, Charles. 1984. *Normal Accidents: Living with High-Risk Technologies*. New York: Basic Books.

Pigeon, Patrick, and Julien Rebotier. 2016. *Disaster Prevention Policies: A Challenging and Critical Outlook*. London: ISTE Publishers, Elsevier.

Primoratz, Igor, ed. 2010. *Terror from the Sky: The Bombing of German Cities in World War II*. New York: Berghahn Books.

Quarantelli E.L., ed. 1998. *What is a disaster? A Dozen Perspectives on the Question*. New York: Routledge.

Rangasami, Amrita. 1987. "Famine: The Anthropological Account." *Economic and Political Weekly* 22, no. 33, August 15, 1987. Mumbai.

Remes, Jacob A.C. 2016. *Disaster Citizenship; Survivors, Solidarity, and Power in the Progressive Era*. Champaign IL: University of Illinois Press.

Roy, Arundhati. 2019 *My Seditious Heart, Collected Non-Fiction.* Toronto: Hamish Hamilton (imprint of Penguin).
Schmidt, Klaus. 1964. *Die Brandnacht: Dokumente von der Zerstörung Darmstadt am 11. September, 1944.* Darmstadt: Reba-Verlag.
Schuilenburg, Marc. 2012. "The Securitization of Society: On the Rise of Quasi-Criminal Law and Selective Exclusion." *Social Justice*, Special issue: Policing the Crisis-Policing in Crisis, 38 no. 1:273–278.
Sliwinski, Alicia. 2018. *A House of One's Own; the Moral Economy of Post-Disaster Aid in El Salvador.* Montreal: McGill-Queen's University Press.
Sorokin, Pitirim A. 1942. *Man and Society in Calamity: The Effects of War, Revolution, Famine and Pestilence Upon Human Mind, Behavior, Social Organisation and Cultural Life.* New York: E.P. Dutton.
Standing, Guy. 2011. *The Precariat: The New Dangerous Class.* London: Bloomsbury Academic.
Steinberg, Ted. 2000. *Acts of God: The Unnatural History of Natural Disaster in America.* 2nd edition. New York: Oxford University Press.
Stockholm International Peace Research Institute. 2021. *SIPRI Yearbook 2021: Armaments, Disarmament and International Security.* Oxford: Oxford University Press.
Strolovitch, Dara. 2021. "When Does a Crisis Begin? Race, Gender, and the Subprime Noncrisis of the Late 1990s". In *Critical Disaster Studies*, edited by Jacob Remes, and Andy Horowitz. Philadelphia: University of Pennsylvania Press.
Sudmeier-Rieux, Karen, Manuela Fernández, Ivanna M. Penna, Michel Jaboyedoff, and J.C. Gaillard, eds. 2017. *Emerging Issues in Disaster Risk Reduction, Migration, Climate Change and Sustainable Development.* Heidelberg: Springer.
Svistova, Juliana, and Loretta Pyles. 2018. *Production of Disaster and Recovery in Post-Earthquake Haiti.* London: Routledge.
Trades Union Congress. 2008. *Hard Work, Hidden Lives. The Full Report of the Commission on Vulnerable Employment.* https://www.tuc.org.uk/publications/hard-work-hidden-lives-short-report-commission-vulnerable-employment
Trimikliniotis, Nicos. 2017. *Combating Undeclared Work as a Work-Related Issue: Protecting Work and Respecting Fundamental Rights.* docx. European Commission, Oslo, Norway: Directorate-General for Employment, Social Affairs and Inclusion.
Turner, Barry A. 1978. *Man-Made Disasters.* London: Wykeham Publications.
United Nations Development Programme, Bureau of Crisis Prevention and Management. 2004. *A Global Report, Reducing Disaster Risk: A Challenge for Development.* New York: UNDR.
United Nations Office for Disaster Risk Reduction. 2020. *Prevention Web.* Geneva: UNDRR. https://www.undrr.org
United Nations International Strategy for Disaster Reduction. 2005. *Hyogo Framework for Action 2005–2015.* https://www.unisdr.org/2005/wcdr/intergover/official-doc/L-docs/Hyogo-framework-for-action-english
United Nations International Strategy for Disaster Reduction. 2010. *Natural Hazards, Unnatural Disasters: The Economics of Effective Prevention.* World Bank. © World Bank. https://openknowledge.worldbank.org/handle/10986/2512 License: CC BY 3.0 IGO. https://www.unisdr.org/we/inform/publications/15136
USSBS E 1945a. *Over-All Report.* United States Strategic Bombing Survey (European War). Office of the Chairman, Report #2. Washington, D.C.
USSBS E 1945b. *The Effect of Bombing on Medical and Health Care in Germany.* United States Strategic Bombing Survey (European War). Medical Branch, Morale Division, Report #65. Washington, D.C.

USSBS E 1947a. *Civilian Defense Division Final Report.* United States Strategic Bombing Survey (European War). Civilian Defence Division, Report #40. Washington, D.C.

USSBS E 1947b. *The Effects of Strategic Bombing on German Morale.* United States Strategic Bombing Survey (European War). Morale Division, Report #64b. Washington, D.C.

USSBS P 1947a. *The Effects of Strategic Bombing on Japanese Morale.* United States Strategic Bombing Survey (Pacific War). Morale Division, Report #14. Washington, D.C.

USSBS P 1947b. *The Effects of Bombing on Health and Medical Services in Japan*, Medical Division, Report #12. Washington, D.C.

Virilio, Paul. 2007. *The Original Accident.* Translated by Julie Rose. London: Polity Press.

Waite, Louise. 2009. "A Place and Space for a Critical Geography of Precarity?" *Geography Compass* 3, no. 1: 412–433.

Watts, Michael J. 1983. *Silent Violence: Food, Famine, and Peasantry in Northern Nigeria.* Berkeley: University of California Press.

White, Gilbert Fowler, ed. 1961. *Papers on Flood Problems.* Department of Geography, University of Chicago, Research Paper 70. Chicago: University of Chicago Press.

Wildavsky, Aaron B. 1981. "Richer is Safer." *Financial Analysts Journal* 37, no. 2 (April): 19–22. https://www.jstor.org/stable/4478435

Wisner, Ben. 2012. "Violent Conflict, Natural Hazards and Disaster". In *The Routledge Handbook of Hazards and Disaster Risk Reduction*, edited by Ben Wisner, J.C. Gaillard, and Ilan Kelman, 71–81. London: Routledge.

4

CREATING DISASTER RISK AND CONSTRUCTING GENDERED VULNERABILITY

Sarah Bradshaw, Brian Linneker, and Lisa Overton

Introduction

The economic growth focus of the 'modernisation' development discourse and related policy and practice has not only harmed the planet and people but also created many of the 'disasters' policy makers then seek to mitigate. In the modernisation discourse economic growth was originally viewed as being 'gender neutral' and this meant women were largely excluded from the development narrative (Jaquette 1982). While current development policy and practice includes women, they are targeted as 'deliverers' rather than beneficiaries of projects, highlighting women's inclusion can be as problematic as their exclusion (Bradshaw 2014). Taking as its starting point that the development model creates disasters, this chapter explores the extent to which development also creates a specific gendered vulnerability to disaster.

Early conceptualisations of disasters constructed them as the same as the 'natural' hazard that caused them (Cardona 2004) and this focus on the 'natural' often rendered them 'neutral' in gender terms (Fordham 1998). The shift to a vulnerability focussed disaster discourse recognised that differences and inequalities between people largely determine the outcomes of a hazard event. Within this, gender roles and relations are important in understanding the differential impacts of hazards (Blaikie et al. 1994). While the notion of vulnerability sought to reflect diversity of experience, there were problems with how it came to be applied by policy makers, who often use it to construct whole countries, regions, and groups of people as 'vulnerable'. This conceptualisation allows other countries and actors to intervene, justifying their actions in the name of 'development' and in the post-disaster context as 'humanitarianism' (Bankoff 2001). Longer term post-disaster interventions are constructed as positive, aiming to 'build back

DOI: 10.4324/9781003219453-5

better' the physical and social infrastructure and through this reduce vulnerability (FEMA 2000).

Within the vulnerability discourse women were initially constructed as victims of disaster. While the recent shift to notions of disaster 'resilience' has seen women and girls re-created as a 'resource' for disaster risk reduction (Bradshaw et al. 2020) societal norms still construct women as a specific group in need of help in times of crisis. Positioning Southern women's experiences and interests against a heterosexual, White, middle-class, Western women reference point (Mohanty 2003) means in the Global South women are often pictured as 'poor, pregnant and powerless' (Win 2009) with post-disaster bringing a 'window of opportunity' to address women's vulnerability (Byrne and Baden 1995). This chapter begins by examining the basis for assuming women's vulnerability. From the outset it should be noted the focus of the chapter is on women as a group and as they relate to men, but this is not because we see gender as a binary nor because we understand women to be a homogenous group. It is because, as the chapter will discuss, the existing evidence base for policy interventions and the dominant policy discourse is binary, and focussed on women. It pays little attention to intersectionality or differences between women, and instead assumes commonalities exist among women, including an assumed shared vulnerability.

Assuming women's vulnerability

While 'gender' is the terminology used in mainstream policy discourse, a focus on differences based on sex or biology remains dominant, and the focus on women as a category for actions and analysis continues. Biological differences between men and women that underpin notions such as the 'sexual division of labour' have been demonstrated to have less to do with biology than with socially constructed notions of what it means to be a man or a woman. However, the notion that hormonal and physiological characteristics render women 'weaker' than men remains a strong social narrative. Carpenter (2005) when discussing the constructed vulnerability of women in the context of conflict, notes that at least since the Middle Ages the discourse of women as the weaker sex has dominated. This supposed physiological 'weakness' is implicit in the Geneva Convention when suggesting that women shall be 'treated with all consideration due to their sex' (Gardam and Jervis 2001, 95), while the 1973 Declaration on the Protection of Women and Children in Armed Conflict states explicitly that 'women and children … are the most vulnerable members of the population'. As with conflict, a disaster is a violent act, and as women are assumed to be the 'weaker sex' so they are assumed to be more impacted, with the UN suggesting 'women always tend to suffer most from the impact of disasters' (UN/ADPC 2010, 8).

Of course, not all women are weaker than all men, and many women are stronger and quicker than their male counterparts (especially younger women compared to elderly men). However, images of women used by NGOs and others post-disaster to raise money, often reinforce the idea of women as hapless, crying victims. Women are often pictured as unable to properly care for children due to the 'disaster', yet paradoxically also presented as the means to bring positive change if only their 'natural' mothering skills could be realised. Thus, a specific normative woman is created and visibilised at the expense of actual women who are diverse in their identities and experiences. In the dominant discourse, biology is constructed as ahistorical, unchanging, and unchallengeable when in fact, feminist and queer scholarship has demonstrated that the concept 'sex' is just as constructed as gender (Butler 1990). This socially constructed homogenised view of women as weak is justified through a biology that is based on women's childbearing and 'related' childrearing roles. Societies that see the main value and role of women as mothers will often subconsciously when thinking of women think of them as pregnant, and when not pregnant encumbered by children. Heavily pregnant women or a woman with a small child to carry may indeed be slower and less able to respond to a fast-moving and sudden onset hazardous event (the stereotypical notion of 'disaster'). The unconscious image of all women as mothers or potential mothers helps explain how all women are then constructed as 'vulnerable'. When women are too young to be mothers they are seen to be 'children' and when seniors, are seen as weak and frail, and rendered vulnerable by their old age. At all life stages, women are continually and artificially seen to be 'naturally' weaker and thus their vulnerability is assumed.

Even when the discourse is more nuanced and women's vulnerability is linked to economic, social, and political characteristics and inequalities, there is still an issue with the assumptions being made about women and men. This includes a lack of gender data and evidence to even support the notion there is a feminised impact of disaster. The 'evidence' base includes vague suggestions such as more women than men died during the 2003 European heatwave (Pirard et al. 2005); specific claims such as of the 140,000 people who died in Bangladesh cyclones 90% were women (Ikeda 1995); to the now infamous 'women, boys and girls are 14 times more likely than men to die during a disaster' (attributed to Peterson 2007 cited in WUNRN 2007). The most robust evidence for a 'feminised disaster mortality' lies with analysis by Neumayer and Plümper (2007) which actually concludes that in situations of greater inequality, there is greater chance that more women will die. Thus, it is not sex or biology that determines a feminised outcome, but gender inequality.

That there is a feminised disaster mortality is supported by evidence from the Indian Ocean Tsunamis. However, women's greater vulnerability lay not with their biological 'weakness', but with gendered social codes of conduct that restrict women's bodies and movement, putting them at greater risk than men. In Sri Lanka, gendered vulnerability came from girls not being taught to swim and being told it is not 'proper' for them to climb trees, and from women's saris which wrapped restrictively around their bodies and were a hindrance when there was a need to run fast (Oxfam 2005). Across the globe many women remain in their homes in the face of hazards, such as rising rivers as cultural practices mean they cannot leave home without male agreement or accompaniment (Ariyabandu

2009). In many parts of the world, fathers, husbands, brothers, and sons restrict the mobility of 'their' daughters, wives, sisters, and mothers keeping them 'off the streets' and out of public places, and confining them to the 'safe' space of the home and away from (other) violent men (Chant and McIlwaine 2016). Despite (other) men being constructed as the threat to women, it is women's mobility and bodies that are restricted under patriarchal social relations (on patriarchy see Patil 2013; Walby 1989, 1990). Control comes from policing women's bodies and sexualities, with violence, and the threat of violence, being one of the main ways in which men's authority is exercised and maintained. Where women 'defy' restrictive gender codes, any experience of violence is constructed as their own fault thereby serving to confirm that women need protecting and their actions restricting. Governance regimes and legal systems, education, and religion further institutionalise gendered power imbalances in intimate personal relationships. Patriarchal structures exist at every level of society, and households may be as much sites of oppression as of solidarity for many women (Chant 2007). The household itself can create gendered 'vulnerability' and be a site of differential risk (Bradshaw 2013). The vulnerabilities household relations create are also male vulnerabilities, in as much as societal expectations of men to protect and provide for the family may place them in risky situations (Bradshaw et al. 2017b). Although gendered risk is socially constructed, and thus who will be most at risk can vary, policy makers still work on the basis that women are more 'vulnerable' than men.

Policy makers might promote the 'women as vulnerable' discourse because the alternative – that socially constructed roles determine vulnerability – may be more difficult to put into practice. The discourse fits with societal norms and the use of selective qualitative evidence over quantitative data seems to be acceptable when it supports the dominant policy discourse. The selective use of data by policy makers is also apparent in conflict contexts as Carpenter (2005, 319) notes the narrative of women and children being more likely than men to be the direct targets of attack remains dominant, despite existing date highlighting that civilian men and older boys are most likely to be directly killed in war or civil strife (Jones 2000; Goldstein 2001). The continued focus on women and children as 'victims' she suggests is in part explained by the social acceptability of this focus, and the socially accepted understanding of women and children's inherent vulnerability.

Since vulnerability is not inherent, and women and children are not necessarily more vulnerable, moving away from the 'natural' discourse to the economic and social discourse demands new assumptions are made to maintain the myth of their greater vulnerability.

Although 'the poor' are not always vulnerable as income does not always protect from a hazard event when social networks and local knowledge may be the best protection, in a world where progress is measured by wealth, and development is determined by economic growth, poverty has often become a shorthand for vulnerability. The assumption that poverty determines vulnerability is compounded in gender terms by the assumption that women are poorer than men. It follows then that women are more vulnerable than men due to their greater poverty. While the so-called 'feminisation of poverty' has been contested (Chant 2008a, 2008b) the assumption of women's poverty and by association the assumption of women's vulnerability, still underpins the policy focus on women.

Including women by assumption

The field of gender and disasters, perhaps rather mistakenly, has looked to gender and development for lessons to learn and a path to follow (Bradshaw 2014). Sparked by research in the 1970s that argued that male-led and male-centric development practices harmed women (Boserup 1970), development practitioners looked to address this by 'adding women in' to the existing development approach, rather than question that approach. While this has been critiqued, and other more nuanced approaches developed, the 'women in development' approach remains dominant in institutional practices (see Kabeer 1994 for discussion of the evolution of gender and development thinking). The acceptance of the need to include women in development was illustrated by women having a 'goal of their own' in the 2000 Millennium Development Goals, and women are now not just included in development, but central to the development process (Cornwall and Rivas 2015). Gender critiques then no longer focus on women's exclusion and the need to include women, but rather how women are included and why (Prügl 2017). We might also ask 'by whom'? Roberts (2012) notes governments and corporations, as well as many gender activists and development actors, aim to promote women's equality within a neo-liberal model of development that is inherently inequitable and exploitative. Rather than challenging existing power relations, such gender champions are 'walking the halls of corporate and state power' and appear to have 'gone to bed with capitalism' (Prügl 2015, 614). More generally, feminist scholarship has problematised power relations between women for some time, particularly Black and postcolonial feminists (Yuval-Davis 2006, 1197; Jayawardene 1995; Crenshaw 1989; Mohanty 1988). Multiple positionings of gender, race, class, sexualities, disabilities among others mean that some women are likely to be more impacted by unequal gendered power relations than others. Kimberlé Crenshaw (1989) coined the term 'Intersectionality' to visibilise this issue, which originally referred to the unique experience Black women faced as a result of 'dual oppressions' that could not be explained by either race or gender discrimination on their own. Since its inception the term has been broadened to include other characteristics, and intersectionality offers a way of understanding women's oppression as existing in its endless variety, yet monotonous similarity (Rubin 1975).

Women then are not a homogenous group and it is as vital to understand the differences between women as it is to know the commonalities of inequalities. Yet what do we actually know? Knowledge of gender as a lived reality is largely missing from official discourse but even the most basic level of knowledge of gendered difference, in income, for example, is also often lacking or its evidence base questionable. Women's inclusion in development is often justified on the grounds of women's relative income poverty. The focus on women and income poverty began with the famous claim from the Beijing conference in 1995 that '70% of the world's poor are women and it is rising'. This assertion gave rise to the notion of a (global) 'feminisation of poverty', a notion popularised in part through research by UN agencies (Medeiros and Costa 2008). Relentless repetition in academic publications, policy and programme documents, and social media has constructed this feminised poverty as 'fact'. For example, as late as 2016 the Deputy Director of UN Women suggested that sustainable development is not possible if the 'feminisation of poverty' continues (Puri 2016). Yet in 2015 UN Women's own report on the 'Progress of the World's Women' stated, albeit in a footnote, that 'the much cited "factoid" that 70% of the world's poor are women is now widely regarded as improbable' (UN Women 2015, 307, 92n) and noted in a box on poverty that 'it is unknown how many of those living in poverty are women and girls' (UN Women 2015, 45, Box 1.4).

Fifty years since Boserup suggested that development harmed women and over 20 years since poverty was constructed as feminised, we still do not have a nuanced understanding of women's relative deprivation, and the only thing we can say with certainty is 'we don't know' (Bradshaw et al. 2017a, 2019a). But why don't we know more about the situation and position of different women? It may be true that 70% of the world's poor are women, equally it could be entirely false and the only way to know is to fill the gender data gap. This begs the question why have we been relying on unreliable statistics for so long? It may be because it is too difficult to know, as we need to interview men and women separately inside a diverse range of households. While timely and costly, such studies are beginning to be undertaken proving this can be done on a large scale (see Wisor et al. 2014). It could be argued it is because there is no political will to fill the gender data gap, pointing to the World Bank, UN and national governments as not prioritising gender enough to want to find out. However, it might also be argued that mainstream policy makers don't actually want to know more about gender issues, as they already know all that they want to know. Gendered data might unsettle their assumptions about women, and women in the Global South in particular. Global South women are useful to policy makers, and they are useful because of their assumed poverty, and vulnerability, not despite this.

Feminising policy

One question to address seems to be why is women's assumed poverty and vulnerability so useful for policy makers? In 1996 Cecile Jackson wrote an article with the title 'Rescuing Gender from the Poverty Trap'. Her argument was that the idea that poverty has a 'female face' meant that any development policy and projects around poverty could be said to be doing work to improve the situation of women (as women are 'poorer') and so doing poverty was 'doing gender'. On the other hand, early gender projects often took the form of adding women into education and employment, both a means to improve economically so these projects were also 'doing poverty'. Poverty and gender had then become inexorably linked and 'gender' had become something done to women to achieve other aims.

Actors such as the World Bank came to understand that when women and girls have systematically lower access to education and health this translates into 'less than optimal' levels of labour market participation and entrepreneurship (World Bank 2001; WBGDG 2003). Saying women are 'poor' gives a justification for targeting women with resources, resources that are not aimed at addressing women's poverty, but rather poverty more generally, and in particular, because of the assumption that all women are or will be mothers, that of children. At the same time, it constructs inequality as a 'women's issue' rather than conceptualising it within a patriarchal and capitalist system that is essentially hierarchical and exploitative. In practical terms this is perhaps not surprising as addressing gender inequality through addressing 'women's' poverty is doable and acceptable for development actors while addressing structural inequalities of power is not.

The feminisation of poverty discourse translated into a feminisation of poverty alleviation. Women are targeted by policies and projects not for being 'poorer' but for being mothers, and their 'naturally' more caring nature means when income and resources enter the household via women the wellbeing of those in the household improves. The socially constructed 'good' behaviour of women is in contrast to the 'irresponsible' behaviour of men which sees men withhold income for their own personal use as they act out the social constructions around what it means to be a man, and how men perform masculinity (Bradshaw et al. 2017b). In practical terms, these social conventions mean, for example, that reduced child mortality risks associated with increases in household income are almost 20 times as large if the income is in the hands of the mother rather than the father (WBGDG 2003). Giving resources to women then is an efficient use of resources. In the disaster context we can see this same logic and women are targeted as virtuous-victims post-event (Bradshaw 2013), the targeting of women justified through their vulnerability making them the 'victims' of events, but also because they are 'virtuous', using resources to reduce the vulnerability of others.

However, just 'giving' resources to people, even virtuous women, goes somewhat against the logic of actors such as the World Bank and so cash transfers often come with conditions. These conditions are explicitly focussed on children – increasing their health and education – and the responsibility to meet them explicitly that of women – as the 'beneficiaries' of the cash. There are added burdens in time spent 'walking and waiting' (Cookson 2016) as well as attending clinics and workshops and/or the opportunity cost of money lost from engagement in income generation activities (Bradshaw 2008). This has led to the suggestion that dealing with poverty is 'arguably as onerous and exploitative as suffering poverty' and as involvement in such programmes has become less negotiable over time there has been a 'feminisation of responsibility and/or obligation' (Chant 2007, 2008b, 2016).

This 'feminisation of responsibility' is not confined to poverty alleviation programmes but is also clearly seen in post-disaster relief and reconstruction projects. Women have become a target for reconstruction projects in recent years, but not necessarily to promote equality or for strategic gains, but because of efficiency in practical terms (Bradshaw and Linneker 2009). An in-depth study of four communities in Nicaragua impacted by hurricane Mitch (Bradshaw 2002) suggested that while half the women interviewed perceived that they participated most in the projects for reconstruction, only a quarter felt that it is women who benefit from this 'participation'. The majority stated that it was the family that benefited from their participation in reconstruction. Such perceptions were supported by interviews with representatives from a number of NGO organisations that targeted resources at women, including 'non-traditional' resources such as giving women collective ownership of cows. For example, when asked how men had reacted to this focus on women one representative commented that there had not been any major problems since 'the women have their cows and the men are drinking the milk…'. Thus while portrayed as a project focussed on women and promoting gender equality, in fact the outcome might bring more work for women, and greater benefits to men.

Far from challenging women's stereotypical roles, reconstruction initiatives may then reinforce these, making response to disasters another element of women's caring role, and increasingly disaster risk reduction is also being seen to be better undertaken by women. The notion that women are 'closer to nature' through their birth-giving role is also exploited and mainstream gender debates in development continue to construct the 'Global South Woman' to be even closer to her biology (Koffman and Gill 2013; Wilson 2015). This supposed natural affinity with nature means increasingly caring for the planet is constructed as a woman's responsibility. All this expands what women as mothers 'should' do as part of 'their' role to include poverty alleviation, disaster risk reduction, and climate change adaptation.

Being good, naturally

While women are targeted post-event to deliver goods and services to others in the name of 'gender equality', they are also increasingly targeted prior-event in activities linked to environmental protection and climate change mitigation. Discussion of the 'feminisation' (Chant 2008b; Bradshaw 2010) or 'motherisation' (Molyneux 2006) of policy response, or an 'ecomaternalism' (Arora-Jonsson 2011) is not new within gender and environmental literature. The Women Environment and Development (WED) approach emerged in the 1990s and has at its base notions of 'ecofeminism' which tended to promote the 'women as closer to nature' discourse, prioritising biology as the explanatory factor for women's supposed natural affinity with nature, an approach critiqued as not only presenting women as a homogenous group but also as essentialising women (Leach 2007). Actors such as the World Bank were able to use this discourse to suggest a 'win-win' approach to environment and gender (see Jackson 1998) that saw women constructed as 'chief victim-and caretaker' (Resurreccion 2012) appropriating women's unpaid labour in activities to protect the environment.

Later approaches more clearly recognised the sexual (or better stated gendered) division of labour as influencing the supposed closeness of women to nature and the impact of wider social and political processes, in that often 'closeness' comes not naturally but through necessity, for example from knowledge gained through having to farm the most marginal lands. Yet biology still tends to dominate the 'mother earth' discourse and it is interesting that while the climate change discourse is very 'masculinised', framed in scientific language, and developed within a securitised agenda, the solution to the problem is often presented as lying with women's 'natural' affinity to nature (see Alston and Whittenbury 2013; Cela et al. 2013; Dankelman 2010). As such, care needs to be taken that we do not construct all women as 'naturally' more caring of the environment and thus suggest they are the ones that should be charged with its protection (MacGregor 2010). Saying women are 'vulnerable' to hazards because they are poor and because they are women (and because they are in the Global South), and at the same time highlighting how they are 'naturally' closer to nature means they can be the target of projects to reduce their (constructed) vulnerability, but that really aim to address wider environmental concerns. Women have been constructed as the protectors of the environment and are now being constructed as those who can mitigate the outcomes of environmental hazards and reduce disaster risk. Yet paradoxically they are also constructed as the ones needing protection during and after disaster.

Protecting the protectors

In the gender vulnerability discourse women are constructed as vulnerable because they are women and as women they need protecting. It almost 'goes without saying' they are vulnerable from their biology, and it is this 'biology' that makes them vulnerable in another way, or rather it is 'biological' attraction that makes them vulnerable. In this parallel biological discourse women are in danger – the danger here is not disaster, but men and men's 'biological urges' which are also often constructed as something 'natural' and 'uncontrollable'. The notion that men 'out there' are waiting to harm women if they leave the 'safety' of the home is a common discourse used to control women's mobility and justify men accompanying women when they do leave the home. If women are constructed as needing to be protected in everyday life or 'normal' times, when 'normality' is disrupted there is an even greater (constructed) need to protect women. This protectionist discourse in conflict situations sees actions such as UN resolutions presented as safeguarding women and children yet using the safety and protection of women as justification for (violent) intervention (Otto 2016).

It is important when considering violence to move away from focusing on individual men or groups of men as 'the problem' and instead focus on the structures and systems that utilise violence to allow men to dominate, or the notion of patriarchy. It is important also to move away from simplistic understandings of patriarchy as male control over women and to focus on the related ideologies of masculinities and femininities it produces. Hierarchical male-male relations of inequality are equally as important as male-female relations, as patriarchy may be defined as a set of social relations between men which, although hierarchical, establishes an interdependence and solidarity between them which allows them to dominate women (Hartmann 1981, 14). On a day-to-day basis, these hierarchies exist along class, age, and race lines, among others, and may be intensified in 'non-normal' contexts. In conflict contexts, Henry (2017) has used the notion of 'clean masculinity' to highlight how some men (the brave and selfless peacekeepers) are positioned hierarchically in relation to 'native' (read: violent) masculinity and this sets White men as the saviours of 'Black and Brown women.' Equally female 'saviours' may oppress (other) women. Western 'feminists' in 'going to bed with capitalism' (Prügl 2015) may also reaffirm the 'patriarchal bargain' (Kandiyoti 1988) where female-female hierarchal relations help to maintain patriarchal systems. Systems that reward women for policing and enforcing patriarchal power, troubling the already problematic notion of a sisterhood between women (Henry 2012).

The notions of protecting women, or better stated, the policing of women's mobility and bodies are central to maintaining patriarchal systems, and violence and the threat of violence is the justification and means to do so. For example, post-Haiti earthquake, the World Food Programme allocated family food vouchers solely to women at distribution points recognising that access to food can change power relations and was seen as a means to protect women from violence (cited in Uwantege Hart 2011). However, by 'banning' men from entering the distribution points, the surrounding areas became risky with reports of women being robbed and even coerced into sex for extra coupons by the male military escorts who were meant to protect them. De Alwis' (2016) post-tsunami research suggests that such pathologisation of men as violent, particularly where the discourse is ethnicised and linked to the so-called 'inescapable' legacies of violent conflict, has had devastating impacts on surviving men. If men see themselves as having 'failed' in their socially constructed role as protector, and if living off aid having 'failed' in their socially constructed role as provider, with this 'failure' reinforced by relief actions that target women with resources, then there is potential for amplifying and intensifying patriarchal attitudes and actions, including violence (Bradshaw 2013). Men may compensate for their loss of ability to fulfil socially-prescribed roles of protector and provider by (re)asserting authority and power through Gender Based Violence, both male-on-male as well as Violence against Women and Girls (VAWG), and in its most extreme form through the rising number of cases of femicide – the violent and deliberate killing of women on grounds of their gender – across the globe (Bradshaw et al. 2019b). However, while VAWG is now assumed post-disaster, it is constructed as abnormal and disaster-associated rather than an intensification of 'normal' everyday reality and rather than measured and monitored over time, it is assumed to increase (Bradshaw and Fordham 2013). This leads to interventions to protect women against violence in the specific context of disaster. As conflict research has highlighted, this may lead to an overemphasis on violence that can only be understood by security specialists, rather than as a social reality (Richards 2005). It also constructs violence as something to be addressed by security specialists.

In the Global South, 'protection' of women in conflict or disaster traditionally has come in the form of heroic Western military personnel and male relief workers, who fly into dangerous places to save vulnerable women (Razack 2004). In the contemporary conflict context, the majority of those keeping the peace in the Global South are now peacekeeping forces drawn from the Global South, with an increasing number of women peacekeepers in these missions (Henry 2012). However, post-disaster, Western governments continue to send military and Western aid workers to 'help' in relief and recovery efforts. Interventions are justified as protection, and such heroic actions demand gratitude. As gender intersects with a highly racialised set of relations embedded in colonial gratitude, what gains access to aid are feminine not masculine characteristics, showing weakness gains access to resources, showing gratitude is the correct response once access is gained. (Razack 2004; Zisk-Marten 2004). Women must enhance their victim status to enable their access to aid, what Hilhorst et al. (2018, 58) conceptualise as a 'strategic essentialism' or what Utas (2005) labels as 'victimcy'. While the original patriarchal bargain with male (partner or relative) as protector and provider is disrupted at times of disaster and more generally and slowly by perceived feminised 'gains', women are encouraged to enter into new patriarchal bargains with new protectors and providers, with what Hilhorst (2016) suggests is a complementary enactment of 'ignorancy' by aid workers who play along with this representation of women as victims in order to secure the continued provision of aid.

Revelations of sexual abuse and assault among (White, Western) aid workers highlight that women may actually need protecting from their 'protectors'. Despite reports emerging since at least 2008 that aid workers were sexually exploiting disaster survivors, it was not until almost 10 years later with the emergence of a sub-tag of #MeToo that this came to mainstream attention. At the centre of this scandal was Oxfam's Haiti response revealed by the now award-winning whistle-blower, Helen Evans. The results saw a number of 'humanitarian' NGOs issue statements and policies as public outcry sparked by #AidToo called for aid funding to be cut. Yet even when accepting there is a problem the discourse of aid organisations is still that the problem is with 'some men', and that we should not let the actions of the few cloud the good work of the many. This institutional response to allegations of gross misconduct sees the responsibility of rights abuses relocated 'away from national politics, international duty-bearers, and from the agencies as proxy authorities, towards individual men as violators of women's rights' (Hilhorst and Jansen 2012, 901). In contrast, when these aid organisations move in to intervene post-disaster in other countries they take the actions of the few and suggest they are indicative of the many, further justifying their interventions and actions. If disasters intensify already destabilised male roles through playing on a feminised vulnerability, targeting women with resources and intervening to 'protect' them, the result may be a 'patriarchal push-back' in the form of individual and collective violence in response to perceived, if not real, advances for women and girls (Bradshaw et al. 2019b). The actions of protection in themselves may then be constructing the need to protect.

In her analysis of the violence witnessed in the post-2011 'Arab Spring' in North Africa and the Middle East, Kandiyoti (2013) questions whether (re)-assertions of male power should be seen as 'patriarchy in action' or 'patriarchy in crisis', as the 'resurgence of traditionalism' with an intensification of (traditional) expressions of masculinity, or as a new form of asserting patriarchal control. Exaggerated and violent expressions of masculinity can be read as 'abnormal' or as an intensification of the 'normal', what Bradshaw et al. (2017b) have referred to as 'supernormal patriarchy'. This notion, developed in relation to the extractive industries in the Global South, highlights that the (hierarchical) relationships (of inter-dependency) formed in the context of 'danger' are not 'abnormal' but an intensified 'normal'. Just as the impact of disasters often produce only a 'more acute, more extreme' form of the daily suffering (Cannon 1994) and given disasters reveal 'existing national, regional and global power structures, as well as power relations within intimate relations' (Enarson and Morrow 1998) then the violence and vulnerability associated with disasters should not be read as ab-normal but as super-normal.

Concluding thoughts: on the natural and the normal

Disasters and related constructs such as gendered risk and vulnerability are not natural but are to be expected as 'normal' outcomes of the economic growth model of the neo-liberal era. Disasters reflect and intensify rather than disrupt this normality, leading not to ab-normal but rather super-normal experiences of everyday realities. Constructing women as 'naturally' vulnerable in the everyday negates the need to recognise that vulnerability is constructed, and instead allows policy makers to be seen to address women's vulnerability via addressing 'their' needs related to biological sex, as mothers. The resilience discourse while seemingly constructing women as 'non-vulnerable' still plays on women's 'natural' attributes, and constructs them as a resource, an efficient deliverer of services to others, playing again on women's mothering role.

The non-natural gendered discourse is confined to 'non-normal' contexts such as post-disaster, when power structures are revealed and played out in the public. In order to explain the violence and inequalities revealed they are constructed as the 'outcomes' of disasters, and so high levels of violence against women are constructed as abnormal and disaster related rather than a revealed or intensified normality. At the same time this allows (some) men to protect (other) women from this assumed new or 'non-normal' violence. As male-male hierarchical relations are part of existing patriarchal structures, these acts of 'protection' may promote 'patriarchal-pushback' which construct the need to protect women, not only reinforcing the woman-as-vulnerable discourse, but creating women's vulnerability.

Women then are not naturally vulnerable to disasters, nor are disasters natural, but the discourse and practice of development creates disasters, and while development constructs women as vulnerable, disaster or rather disaster response, creates women's vulnerability.

References

Alston, Margaret and Whittenbury, Kerri (eds.) 2013. *Research, Action and Policy: Addressing the Gendered Impacts of Climate Change*, ISBN: 978-94-007-5517-8 (Print) 978-94-007-5518-5 (Online), Netherlands: Springer.

Ariyabandu, Madhavi Malalgoda. 2009. "Sex, gender and gender relations in disasters", in Elaine Enarson and P.G. Dhar Chakrabarti (eds) *Women, Gender and Disaster: Global Issues and Initiatives*, Los Angeles: SAGE Publications India Pvt Ltd, 5–17.

Arora-Jonsson, Seema. 2011. "Virtue and vulnerability: Discourses on women, gender and climate change", *Global Environmental Change*, 21:2, 744–751.

Bankoff, Greg. 2001. "Rendering the world unsafe: "Vulnerability" as Western Discourse", *Disasters*, 25:1, 19–35.

Blaikie, Piers, Terry, Cannon, Ian, Davis, and Ben Wisner. 1994. *At Risk: Natural Hazards, People's Vulnerability, and Disasters*, (also published in 2003 with these authors listed: Ben Wisner, Piers Blaikie, Terry Cannon, Ian Davis), London, New York: Routledge.

Boserup, Ester. 1970. *Woman's Role in Economic Development*, New York: St Martin's.

Bradshaw, Sarah. 2002. *Gendered Poverties and Power Relations*, Managua: Fundación Puntos de Encuentro.

Bradshaw, Sarah. 2008. "From structural adjustment to social adjustment: A gendered analysis of conditional cash transfer programmes in Mexico and Nicaragua", *Global Social Policy*, 8:1, 188–207.

Bradshaw, Sarah. 2010. "Feminisation or De-Feminisation? Gendered experiences of poverty post-disaster", in Sylvia Chant (ed.) *International Handbook on Gender and Poverty*, Cheltenham/Northampton: Edward Elgar, 627–632.

Bradshaw, Sarah. 2013. *Gender, Development and Disasters*, Cheltenham/Northampton: Edward Elgar, ISBN: 978 1 84980 446 2.

Bradshaw, Sarah. 2014. "Engendering development and disasters", *Disasters, Special Issue: Building Resilience to Disasters Post-2015*, 39:1, 54–75.

Bradshaw, Sarah, Chant, Sylvia, and Linneker, Brian. 2017a. "Gender and poverty: What we know, don't know, and need to know for Agenda 2030", *Gender, Place and Culture*, Published online 13 November, 24:12, 1667–1688.

Bradshaw, Sarah, Chant, Sylvia, and Linneker, Brian. 2019a. "Challenges and changes in gendered poverty: The feminization, de-feminization and re-feminization of poverty in Latin America", *Feminist Economics*, 25:1, 119–144.

Bradshaw, Sarah, Chant, Sylvia, and Linneker, Brian. 2019b. "Gender, poverty and anti-poverty policy in Latin America: Cautions and concerns in a context of multiple feminisations and 'Patriarchal Pushback'", in Julie Cupples, Marcela Palomino-Schalsa and Manuel Prieto (eds.) *The Routledge Handbook of Latin American Development*, London: Routledge.

Bradshaw, Sarah and Fordham, Maureen. 2013. *Women and Girls in Disasters*. report produced for Department for International Development, UK, August 2013. https://www.gov.uk/government/uploads/system/uploads/attachment_data/file/236656/women-girls-disasters.pdf

Bradshaw, Sarah, Hawthorne, Helen, and Linneker, Brian. 2020. "The discord between discourse and data in engendering resilience building for sustainability", *International Journal of Disaster Risk Reduction*, 50:101860, 1–7. ISSN 2212-4209 [Article] doi:10.1016/j.ijdrr.2020.101860

Bradshaw, Sarah and Linneker, Brian. 2009. "Gender perspectives on disaster reconstruction in nicaragua: Reconstructing roles and relations?", in E. Enarson and D. Chakrabarti (eds.) *Women, Gender and Disaster: Global Issues and Initiatives*, India: Sage, 75–88.

Bradshaw, Sarah, Linneker, Brian, and Overton, Lisa. 2017b. "Understanding extractive industries as sites of "Super Normal" profits and "Super-Normal" Patriarchy", *Gender and Development*, 25:3, Special Issue on Natural Resource Justice, 439–454. https://www.tandfonline.com/doi/pdf/10.1080/13552074.2017.1379780

Butler, Judith. 1990. *Gender Trouble: Feminism and the Subversion of Identity* (second edition 1999), London: Routledge.

Byrne, Bridget and Baden, Sally. 1995. *Gender, Emergencies and Humanitarian Assistance* (Vol. 33), Brighton, UK: Institute of Development Studies.

Cannon, Terry. 1994. "Vulnerability analysis and the explanation of 'Natural' disasters", in *Disasters, Development and Environment*, A. Varley (ed.) Chichester, New York, Brisbane, Toronto and Singapore: John Wiley & Sons, 13–29.

Cardona, Omar D. 2004. "The need for rethinking the concepts of vulnerability and risk from a holistic perspective: A necessary review and criticism for effective risk management", in Greg Bankoff, Georg Frerks and Dorothea Hilhorst (eds.), *Mapping Vulnerability: Disasters, Development and People*, London: Earthscan, 37–51.

Carpenter, Charli. 2005. ""Women, children and other vulnerable groups": Gender, strategic frames and the protection of civilians as a transnational issue", *International Studies Quarterly*, 49, 295–334.

Cela, Blerta, Dankelman, Irene, and Stern, Jeffrey (eds.) 2013. *Powerful Synergies: Gender Equality, Economic Development and Environmental Sustainability*, UNDP. http://www.undp.org/content/undp/en/home/librarypage/womens-empowerment/powerful-synergies/

Chant, Sylvia. 2007. *Gender, Generation and Poverty: Exploring the 'Feminisation of Poverty' in Africa, Asia and Latin America*, Cheltenham: Edward Elgar.

Chant, Sylvia. 2008a. "Dangerous equations? How female-headed households became the poorest of the poor: Causes, consequences and cautions", in J. Momsen (ed.) *Gender and Development: Critical Concepts in Development Studies*, London, UK: Routledge.

Chant, Sylvia. 2008b. "The 'feminisation of poverty' and the 'feminisation' of anti-poverty programmes: Room for revision?", *Journal of Development Studies*, 44:2, 165–197.

Chant, Sylvia. 2016. "Women, girls and world poverty: Equality, empowerment or essentialism?", *International Development Planning Review*, 38:1, 1–24.

Chant, Sylvia and McIlwaine, Cathy. 2016. *Cities, Slums and Gender in the Global South: Towards a Feminised Urban Future*, London: Routledge.

Cookson, Tara Patricia. 2016. "Working for inclusion? Conditional cash transfers, rural women, and the reproduction of inequality", *Antipode*, 48:5, 1187–1205.

Cornwall, Andrea and Rivas, Althea-Maria. 2015. "From 'gender equality and 'women's empowerment' to global justice: Reclaiming a transformative agenda for gender and development", *Third World Quarterly*, 36:2, 396–415.

Crenshaw, Kimberlé Williams. 1989. "Demarginalizing the Intersection of Race and Sex: A Black Feminist Critique of Antidiscrimination Doctrine, Feminist Theory, and Antiracist Politics", *University of Chicago Legal Forum*, 1989, 139–167.

Dankelman, Irene (ed.) 2010. *Gender and Climate Change: An Introduction*, London: Earthscan Publications Ltd.

De Alwis, Malathi. 2016. "The tsunami's wake: Mourning and masculinity in Eastern Sri Lanka", in Elaine Enarson and Bob (ed.) *Men, Masculinities and Disaster*, London: Routledge. 92–102.

Enarson, Elaine and Morrow, Betty. 1998. *The Gendered Terrain of Disasters*, Westport, Connecticut and London: Praeger.

FEMA (Federal Emergency Management Agency). 2000. *Rebuilding for a More Sustainable Future: An Operational Framework*, Washington, DC: FEMA.

Fordham, Maureen. 1998. "Making women visible in disasters: Problematising the private domain", *Journal of Disaster Studies, Policy and Management*, 22:2, 126–143.

Gardam, Judith Gail and Jervis, Michelle J. 2001. *Women, Armed Conflict and International Law*, The Hague: Kluwer Law International.

Goldstein, Joshua S. 2001. *War and Gender*, Cambridge, MA: Cambridge University Press.

Hartmann, Heidi I. 1981. "The unhappy marriage of Marxism and Feminism: Towards a more progressive union", in Lydia Sargent (ed.) *Women and Revolution: A Discussion of the Unhappy Marriage of Marxism and Feminism*, Boston: South End Press, 1–41.

Henry, Marsha. 2012. "Peacexploitation? Interrogating labor hierarchies and global sisterhood among Indian and Uruguayan female peacekeepers", *Globalizations*, 9:1, 15–33.

Henry, Marsha. 2017. "Problematizing military masculinity, intersectionality and male vulnerability in feminist critical military studies", *Critical Military Studies*, 3:2, 182–199.

Hilhorst, Dorothea. 2016. *Aid–Society Relations in Humanitarian Crises and Recovery. Inaugural lecture*, presented on 22 September 2016, in acceptance of the Chair of Humanitarian Aid and Reconstruction at the *Institute of Social Studies* of *Erasmus University Rotterdam*.

Hilhorst, Dorothea and Jansen, Bram J. 2012. "Constructing rights and wrongs in humanitarian aid", *Sociology*, 46:5, 891–905.

Hilhorst, Dorothea, Porter, Holly, and Gordon, Rachel. 2018. "Gender, sexuality, and violence in humanitarian crises", *Disasters*, 42:S1), S3–S16.

Ikeda, Keiko. 1995. "Gender differences in human loss and vulnerability in natural disasters: A case study from Bangladesh", *Indian Journal of Gender Studies*, 2:171–193.

Jackson, Cecile. 1996. "Rescuing gender from the poverty trap", *World Development*, 3, 489–504.

Jackson, Cecile. 1998. "Gender, irrigation, and environment: Arguing for agency", *Agriculture and Human Values*, 15, 4, 313–324.

Jaquette, Jane S. 1982. "Women and modernisation theory", *World Politics*, 1, 267–284.

Jayawardene, Kumari. 1995. *The White Woman's Other Burden: Western Women and South Asia during British Rule*, New York and London: Routledge.

Jones, Adam. 2000. "Gendercide and genocide", *Journal of Genocide Studies*, 2:2, 185–212.

Kabeer, Naila. 1994. *Reversed Realities: Gender Hierarchies in Development Thought*, London: Verso.

Kandiyoti, Deniz. 1988. "Bargaining with patriarchy", *Gender and Society*, 2:3, Special Issue to Honor Jessie Bernard, 2:3, 274–290.

Kandiyoti, Deniz. 2013. "Fear and Fury: Women and post-revolutionary violence", *50:50 Inclusive Democracy*. https://www.opendemocracy.net/5050/deniz-kandiyoti/fear-and-fury-women-and-post-revolutionary-violence.

Koffman, Ofra and Gill, Rosalind. 2013. "The revolution will be Led by a 12-year-old girl': Girl power and global biopolitics", *Development and Change*, 105:1, 83–102.

Leach, Melissa. 2007. "Earth mother myths and other ecofeminist fables: How a strategic notion rose and fell", *Development and Change*, 38:1, 67–85.

MacGregor, Sherilyn. 2010. "A stranger silence still: The need for feminist social research on climate change", *Sociological Review*, 57:Issue Supplement S2, Special Issue: Monograph Series, *Nature, Society and Environmental Crisis*, 124–140.

Medeiros, Marcelo and Costa, Joana. 2008. "Is there a feminisation of poverty in Latin America?", *World Development*, 36:1, 115–127.

Mohanty, Chandra Talpade. 1988. "Under Western eyes: Feminist scholarship and colonial discourses", *Feminist Review*, 30:1, 61–88.

Mohanty, Chandra Talpade. 2003. *Feminism Without Borders: Decolonizing Theory, Practising Solidarity*, Durham and London: Duke University Press.

Molyneux, Maxine. 2006. "Mothers at the service of the new poverty agenda: PROGRESA/ Oportunidades, Mexico's Conditional Transfer Programme", *Journal of Social Policy and Administration*, Special Issue on Latin America, 40:4, 425–449.

Neumayer, Eric and Plümper, Thomas. 2007. "The gendered nature of natural disasters: The impact of catastrophic events on the gender gap in life expectancy, 1981–2002", *Annals of the Association of American Geographers*, 97:3, 551–566.

Otto, Dianne. 2016. Women, Peace and Security: A critical analysis of the Security Council's vision. Working Paper Series from Centre of Women, Peace and Security. http://www.lse.ac.uk/women-peace-security/publications/wps.

Oxfam. 2005. *The Tsunami's Impact on Women*, Oxfam Briefing Note, Oxfam International. https://policy-practice.oxfam.org.uk/publications/the-tsunamis-impact-on-women-115038.

Patil, Vrushali. 2013. "From patriarchy to intersectionality: A transnational feminist assessment of how far we've really come", *Signs: Journal of Women in Culture and Society*, 38:4, 847–867.

Peterson, Kristina. 2007. *Gender Issues in Disaster Responses*, Church World Service Emergency Response Program. www.ncasindia.org/archives/advocacy_internet/disaster/article9.htm.

Pirard, P., Vandentorren, S., Pascal, M., Laaidi, K., and Ledrans, M. 2005. "Summary of the mortality impact assessment of the 2003 heatwave in France", *Euro Surveill*, 10:7/8, 153–156.

Prügl, Elisabeth. 2015. "Neoliberalising Feminism", *New Political Economy*, 20:4, 614–631

Prügl, Elisabeth. 2017. "Neoliberalism with a Feminist face: Crafting a new Hegemony at the world bank", *Feminist Economics*, 23:1, 30–53.

Puri, Lakshmi. 2016. *Sustainable development is not possible if feminization of poverty continues*, Remarks by UN Women Deputy Executive Director Lakshmi Puri at the Opening Ceremony of the 2016 W20 Meeting on 24 May, 2016. http://www.unwomen.org/en/news/stories/2016/5/lakshmi-puri-speech-at-the-opening-ceremony-of-the-2016-w20-meeting.

Razack, Sherene. 2004. *Dark Threats and White Knights: The Somalia Affair, Peacekeeping, and the New Imperialism*, Toronto, Canada: University of Toronto Press.

Resurreccion, Bernadette P. 2012. "The gender and climate debate: More of the same or new ways of doing and thinking?", in Lorraine Elliot and Mely Caballero-Anthony (eds.) *Human Security and Climate Change in Southeast Asia*, London & New York: Routledge, Taylor & Francis Group.

Richards, Paul. 2005. *No War No Peace: An Anthropology of Contemporary Armed Conflicts*, Oxford: James Currey Ltd.

Roberts, Adrienne. 2012. "Financial crisis, financial firms … and financial feminism?", *Socialist Studies*, 8:2, 85–108.

Rubin, Gayle. 1975. "The traffic in women: Notes on the 'Political Economy' of sex", in Rayner R. Reiter (ed.) *Toward an Anthropology of Women*, 157–210. New York: Monthly Review Press.

UN Women (UNW). 2015. *Progress of the World's Women 2015–16: Transforming Economies, Realising Rights*, New York: UN Women. http://progress.unwomen.org/en/2015/pdf/UNW_progressreport.pdf.

UN/ADPC. 2010. *Disaster Proofing The Millennium Development Goals (MDGs)*, UN Millennium Campaign and the Asian Disaster Preparedness Center. http://www.adpc.net/v2007/downloads/2010/oct/mdgproofing.pdf.

Utas, Mats. 2005. "Victimcy, girlfriending, soldiering: Tactic agency in a young woman's social navigation of the Liberian war zone", *Anthropological Quarterly*, 78:2, 403–430.

Uwantege Hart, Sandra. 2011. *Women only: Violence and gendered entitlements in post-quake food distribution in Port-au-Prince, Haiti*. http://graduateinstitute.ch/webdav/site/genre/shared/Genre_docs/Actes_2010/Actes_2010_UwantegeHart.pdf (accessed 14 August 2012).

Walby, Sylvia. 1989. "Theorising patriarchy", *Sociology*, 23, 213–234.

Walby, Sylvia. 1990. *Theorizing Patriarchy*, Cambridge, MA: Basil Blackwell.

WBGDG. 2003. *Gender Equality and the Millennium Development Goals*, World Bank Gender and Development Group – WBGDG, April 2003. https://documents1.worldbank.org/curated/en/307331468762867954/pdf/Gender0MDGs.pdf.

Wilson, Kalpana. 2015. "Towards a radical re-appropriation: Gender, development and neoliberal feminism", *Development and Change*, 46:4. doi: 10.1111/dech.12176.

Win, Everjoice J. 2009. "Not very poor, powerless or pregnant: The African woman forgotten by development", *IDS Bulletin*, 35:4, 61–64.

Wisor, Scott, Bessell, Sharon, Castillo, Fatima, Crawford, Joanne, Donaghue, Kieran, Hunt, Janet, Jaggar, Alison, Liu, Amy, and Pogge, Thomas. 2014. *The Individual Deprivation Measure: A Gender-Sensitive Approach to Poverty Measurement*, Melbourne: International Women's Development Agency Inc. https://www.iwda.org.au/assets/files/IDM-Report-16.02.15_FINAL.pdf.

Women's UN Report Network (WUNRN). 2007. *Reaching Out to Women When Disaster Strikes*. Soroptimist White Paper.

World Bank. 2001. *Social Protection Strategy: From Safety Net to Springboard*, Washington, DC: World Bank.

Yuval-Davis, Nira. 2006. "Intersectionality and feminist politics", *European Journal of Women's Studies*, 13:3, 193–209.

Zisk-Marten, Kimberly. 2004. *Enforcing the Peace: Learning from the Imperial Past*, New York: Columbia University Press.

5

WHAT MUST BE DONE TO RESCUE THE CONCEPT OF VULNERABILITY?

Terry Cannon

The title of this chapter reflects two concerns I have in how the concept vulnerability is being used/abused and possibly under threat, and why it might need to be 'rescued'. Firstly, as with much other terminology (including sustainability, resilience, empowerment, transformation…) it has become a buzzword widely used without clarity of meaning or definition. In relation to disaster risk and climate change, it is often used without precision, applied as a general characteristic of people (or of buildings and infrastructure), and as a post-event label for those who suffer in a disaster (and therefore having no predictive value to assist in reducing vulnerability). Secondly, its use as an explanatory concept (in helping to identify why a hazard becomes a disaster) is becoming less relevant because some organisations and academics who refer to 'vulnerability' are reverting to calling disasters 'natural' (Chmutina and von Meding 2019). This means that the paradigm shift of the past thirty years in which disasters have come to be widely understood in terms of the social construction of vulnerability is under threat, and if the word vulnerability is used it loses its capacity to be used in preparedness.

Where does vulnerability come from?

This heading is deliberately ambiguous – it is about how the word arose in the context of disaster, but also how the 'reality' of vulnerability is understood to be 'caused' by social processes. The term vulnerability is widely used in relation to natural hazard-related disasters, although forty years ago it would not have been very common at all.[1] Its origins are in several strands of research that began to challenge the orthodox accounts that disasters are 'natural' and mostly about the overwhelming power of Nature. One of the earliest of these shifts was Gilbert White's pioneering studies of floods in the USA. He argued that disasters were

DOI: 10.4324/9781003219453-6

happening because people and human systems were located in places that are unsuitable for settlement, and that standard 'structural' means for controlling floods can be inappropriate and lead to worse disasters (AAG 2006; Kates and Burton 2008). Explaining disasters primarily through the focus on the hazards themselves meant they were isolated from this social context of exposure. White's work on how the economy 'constructed' risk by exposing people to floods was a key moment leading to the vulnerability approach. It showed the importance of the role of the economy and politics in determining risk.

This then enabled White's idea of socially-constructed exposure ('living and building in risky places') to be extended into the analysis of how different political, economic, and social factors lead to people being not just exposed but *vulnerable*. What followed was the search for explanations of disasters in terms of differentiated social groups that suffered to a greater or lesser extent because of class, ethnicity, gender, and other socially determined characteristics. Explanations of disasters emerged that were rooted in systems of power that allocated risk unequally to different groups of people. These sought the 'root causes' of disasters in processes and factors that could be quite remote (in space and time) from the actual disaster event (Oliver-Smith 2019). The effects of social structure, geographic location, and even processes happening over long time periods could now be uncovered as contributing to likely harm happening to specific groups of people (e.g. Hewitt 1997; Blaikie et al. 1994; Varley 1994; Oliver-Smith and Hoffman 1999). This work was led by social scientists whose main focus was the 'Third World'/'global South', especially geographers, anthropologists, and development studies researchers. These arguments suggested that disasters (which appeared to have worse impacts in the global South) can be understood as a result of failed development in which vulnerability to hazards is just another expression of the wider characteristics of poverty, ill-health, malnutrition, poor education, and oppression of women.

The understanding of disaster causation can be seen as a series of shifts: firstly away from blaming god, then with the emergence of scientific understanding of geological and climate phenomena a move to 'blaming' Nature. In countries that transitioned to capitalist 'modernity' the natural sciences began the process of explaining geological and climate hazards, and it became unnecessary to rely on supernatural causes (Dynes 2000). But explanations that rely on a role for god/s in causing disasters persist in much of the world, and these cultural perceptions of risk have tended to be missing from institutional and policy assessments of disaster risk reduction (DRR) (IFRC 2014; Krüger et al. 2015). This problem is briefly discussed later because it constitutes a major gap in vulnerability analysis. Another significant gap involves the inadequate treatment of how risks are perceived, and how it is often essential for people to 'live with danger' in order to have a livelihood. These two components (how vulnerability is linked to cultural/religious explanations of disasters, and why many people need to live in hazardous places) are introduced here as key components of the social construction of vulnerability.

Improving the validity and use of vulnerability analysis

Vulnerability can be understood as having a range of seven determining components (related to economic, political, social, and behavioural processes), and the potential to involve seven different types of harm to people in relation to these vulnerability factors. These are discussed below, but before that it is necessary to be clear on how to 'rescue' the vulnerability concept. This needs to be done in three ways:

- In relation to natural hazards (and climate change) vulnerability must be assessed in relation *to something*. Understanding vulnerability as a vague general characteristic is not helpful as a guide to risk reduction: it involves a particular type of risk that has the potential to cause harm if the person is vulnerable to it. It needs to be specific to the type of hazard: a person (or human system) may be vulnerable to one type of hazard and not to another. Each type of hazard interacts with different characteristics of the exposed people. A poor person in a flimsy house may be safer in an earthquake than a richer person in a badly constructed high-rise apartment, while the opposite may be true in a flood. Vulnerability is often used as a vague general characteristic of people who are poor, marginalised or in other ways to be seen as victims. But this is not very helpful in being prepared for each type of hazard and the *specific* risks involved.
- Vulnerability has to be *predictive*: it involves seven types of potential harm that may be experienced in greater or lesser amounts. These can be identified in advance of a given hazard event. Vulnerability analysis can also identify particular groups or locations with large numbers of people that share similar risk profiles, and this can be used to prepare for and prevent a hazard becoming a disaster.
- Some aspects of vulnerability must be identified as being derived from people's own perception of risks, and their capacity (determined by economic and political factors) and *willingness* to take action to reduce vulnerability (related to two factors: cultural and religious attitudes to risk, and need to live in dangerous locations to have access to a livelihood).

Vulnerability is a complex but comprehensible mix of social factors.[2] These derive firstly from systems of power that lead people to be more or less likely to suffer harm (class, ethnicity, gender), and secondly from people's self-defined ways of living with risk that involve culture (why they think a disaster happens) and perceptions of risk (especially needing to live with danger to get a livelihood). These second types of vulnerability components have been significantly neglected by both the supporters of the social construction paradigm, and by mainstream DRR organisations. Cultural and behavioural factors interact with economic, political, and other social factors, but in the early development of the vulnerability paradigm they tended to be overlooked. The emphasise was on political and economic factors that related to systems of power around class, ethnicity, and gender. Cultural and behavioural aspects of risk perception and risk taking were the subject of a large body of research in psychology and anthropology (and more recently behavioural economics) but these have largely failed to be integrated into natural hazard disaster research and policy, perhaps (as with Beck's work) they were mainly focused on technological risks (examples include Slovic 1987; Slovic and Weber 2002; Douglas and Wildavsky 1983; Kahneman and Tversky 1979).

The idea of using the term *vulnerability* in relation to natural hazards emerged from the 1970s onwards to form the basis for a significant shift in disaster studies to the 'social construction' or vulnerability paradigm (for a summary of early authors see Kelman 2010). It was part of a process in which some social scientists shifted the dominant disaster discourse away from the focus on the hazards themselves (Nature and 'natural' disasters) towards much greater emphasis on how people experience greater or lesser harm because of 'social' factors. This shift was symbolised in the change of language and emphasis during and after the 1990s UN 'International Decades of Natural Disaster Reduction' (IDNDR). This was dominated by experts in natural hazards and had not been influenced by the vulnerability paradigm (Schemper 2019; Shaw 2020). During the decade, social scientists and activists who advocated for the 'social construction' vulnerability paradigm criticised the way the decade had been framed. Resistance came from organisations such as *La Red* (The Network) in Latin America, Duryog Nivaran in south Asia, and Periperi in southern Africa along with several organisations in the Philippines. These worked at grassroots and advocated for DRR as a community-based process. At the same time, many international NGOs that had hitherto mostly been involved in disaster emergency *response* started to get involved in DRR as *prevention and preparedness*. They introduced local-level vulnerability assessment and were also critical of top-down and technical-fix approaches favoured by the UN International Decade and more generally by development banks. At the end of the decade, when the UN ISDR (International Secretariat for Disaster Reduction) was set up with headquarters in Geneva, the vulnerability approach had become much more influential in its policy. The World Bank even issued a report titled *Natural Hazards, UnNatural Disasters* (World Bank 2010).

Much of the discussion about vulnerability was focused on the Third World/Global South and DRR was assumed to be mainly about 'poor people in poor countries'. It was informed by the observation that most of the people who suffer harm from natural hazards (and therefore assumed to have high levels of vulnerability) are also poor. This produced a rather inadequate analysis that equated vulnerability with poverty, arising partly from the increasing involvement of NGOs in DRR who championed anti-poverty actions and community-based disaster risk reduction (CBDRR) as their main approach for working with poor *and vulnerable* people.

The shift in analysis of disasters to the 'vulnerability paradigm' was combined with the argument that disasters are not natural but are socially constructed. It argued that a disaster (related to a natural hazard) has three aspects:

- the hazard itself that has the potential to cause harm to people and human systems; this is mostly 'natural', except that some human action can magnify the hazard (especially floods) or even cause it (as with landslides and road building in mountainous areas);
- the entity that may be harmed by the hazard (usually thought of as people and human systems);
- the processes and factors that make the entity more or less likely to be harmed (these typically include class, gender, ethnicity, religion, dis/ability, and age).

Natural scientists (who often specialise on only one particular hazard type) continued to focus mostly on hazards and their impacts, largely ignoring the processes of social construction of the disaster. DRR can be viewed as a contested area in which different paradigms continue to argue for the value of their favoured approach. At international level, there has been a recent resurgence in using natural disasters as an acceptable terminology (e.g. by some academics, the World Bank, aid ministries, and even by UNDRR (the successor to UNISDR, 2015) and the Red Cross Climate Centre (DFID 2018; ICAI 2018; Chmutina and von Meding 2019; RCCC 2007, 2020 and many other webpages). At the World Bank, using the terminology 'natural disaster' is related to funding conditions as revealed in a 2020 email to me from a senior World Bank staff member that

> my choice of word is not an endorsement of the idea that natural disasters cannot be avoided. It's more an acknowledgement that this is the terms decision-makers and policy-makers (my main audience) use. For instance, we have funding sources that can be used only to reduce the risk from "natural disasters". If I do not work on "natural" disasters, I cannot access those resources. Also, I have not seen any alternative: if a hurricane hitting a city cannot be labelled "natural disasters", then what is the right terms?

This is rather unfortunate as there are several studies of the impact of hurricane Katrina on New Orleans that specifically argue it was not a natural disaster, including the book about Hurricane Katrina *There is no such thing as a natural disaster* (Hartman and Squires 2006).

This reversion to 'naturalness' has strangely coincided with increasing overlap of DRR with some forms of adaptation to climate change, where preparedness is focused on extreme events (especially floods and tropical cyclones) and often focuses on large-scale technical interventions. This is ironic since the increased severity and/or frequency of such hazards is directly 'manmade' through global warming. It appears to be happening in parallel with governments and international organisations restoring terminology that emphasises natural hazards and plays down the role of vulnerability analysis. I suggest this is because using vulnerability analysis requires having explanations rooted in political and economic processes which can be caused (or enabled) by governments and development banks. Playing down (or even ignoring) the creation of vulnerability is common among politicians (either out of ignorance or because they want to avoid being blamed for helping to cause vulnerability). It is also a feature of mainstream media where coverage of disasters attracts viewers and readers but tends to be framed around the 'overwhelming power of Nature'. Causes of vulnerability are often ignored unless there are obvious human failings (as with New Orleans and hurricane Katrina). The focus is often on *nature* and its impacts on society, not the causation of vulnerability within society (Cockburn 2011). Reflections on these and other problems of disaster media coverage are dealt with by Greenberg and Scanlon 2016 and Pantti 2018.

What do we mean when we say a disaster has happened?

Vulnerability needs to be understood in terms of the different types of harm that people can experience. Being a 'disaster victim' has a range of at least seven types of possible impacts, each of which can be experienced to a greater or lesser extent. How a hazard of particular severity causes different harms depends on the level of vulnerability to specific aspects of the hazard (can it kill, cause illness, damage fields, and livelihood, prevent going to work to earn money, and so on). In most DRR approaches, reducing mortality is prioritised. However, most of the harm done in a disaster is experienced by those who do not lose their life, but have assets damaged or destroyed, and livelihoods and income disrupted. It is even possible (for example in Bangladesh where recent cyclone warnings and evacuations seem to have been successful) that very few people die, but hundreds of thousands of people's lives are disrupted for years or even forever. Many years after it happened in 1998, thousands of survivors of hurricane Mitch in central America were found not to have recovered (for Nicaragua see Christoplos et al. 2009; for Honduras see Telford et al. 2004). Fifteen years after hurricane Katrina hit New Orleans in 2005 there is still significant disruption and damage to people, their lives, and livelihoods, and some of the supposed post-disaster assistance (e.g. in housing) made poor and black people's problems worse (Horowitz 2017; Rich 2015; Williams 2020).

Understanding the different types of impacts of a hazard is crucial because in much DRR work (especially that of organisations and governments) there is significant imbalance in which of these harms is given priority. When the term 'disaster' is used in relation to natural hazards, the headline concern is always with mortality: DRR policy mainly aims to reduce the number of deaths. Much less is done to consider what happens when people survive, including with the success of warning systems and evacuations. A much deeper treatment of hazard impacts on vulnerable people is needed. What is meant by a disaster therefore involves finding actions (Table 5.1) in the second column that can reduce the harms of the first column *before* the hazard strikes (and realising that each hazard requires different types of action):

TABLE 5.1 Matching DRR actions to reducing the harms from a particular hazard

Type of harm expected from a hazard impact where preparedness is inadequate	*Link to DRR approaches, prior 'development' and base-line status*	*Phase of action/ Absent*
1. Death	Normal priority – focus of policies to provide: Warnings and evacuations Building codes Flood barriers and zoning	Before
2. Injury	Reductions normally a by-product of reducing deaths; Treatment in emergency relief and recovery;	Before Absent
3. Illness (for example from contaminated water in floods; respiratory diseases; grief and mental illness)	Pre-hazard impact: Clean water sources may / may not be in position to resist hazard impacts Post impact: Emergency response in medical help; Good practice in identification and treatment of bodies to help mitigate grief (identification; acceptable burial)	Before/Absent After
4. Hunger and lack of water	Clean water sources pre-installed in hazard zones that are hazard-proof; Mostly dealt with through emergency relief, often to displaced people	Before After
5. Loss of assets for production and home life (land, livestock, savings, important documents, jewellery used as savings, tools, house)	Methods to protect assets rarely considered in DRR approaches as this often requires changes to systems of power that allocate resources (e.g. land, access to water) and income. A few positive examples: cyclone shelters in Bangladesh now include provision for sheltering livestock; some NGO projects have helped people to protect key documents in floods (e.g. identity cards, school certificates	Largely absent Before
6. Disruption or loss of livelihood. This can be a combination of loss of assets and access to work when buildings and infrastructure are damaged or destroyed, or the result of displacement	In most parts of the global South very little is possible to protect assets such as land and access to water as ownership of these is very unequal and a key component of the systems of power; Some infrastructure is designed to be hazard-proof in some countries; Where people are displaced by a hazard (e.g. floods, cyclone) restoration of livelihoods is often impossible.	Absent Before Absent

(Continued)

TABLE 5.1 (Continued)

Type of harm expected from a hazard impact where preparedness is inadequate	*Link to DRR approaches, prior 'development' and base-line status*	*Phase of action/ Absent*
7. Social and geographical displacement and its impacts on family life, mental health, general well-being	Very little preparedness achieved for these problems in most hazards; impacts and recovery can take years or last forever;	Absent

Most governments and organisations involved in DRR accept that they have a duty to reduce mortality and some have incentives to do it (through commitment to humanitarian goals and the potential for international embarrassment). In most cases (e.g. warning systems, building codes) mortality reduction can be done without significant disruption to systems of power, and could be called DRR 1.0 (using the terminology used for software releases). It does not require significant shifts in the benefits that accrue to the powerful through their control of the economy. This is therefore relatively 'easy' to achieve in the sense that DRR does not need to disrupt the arrangements of the status quo that benefit the powerful.

However, there is generally much less emphasis on what happens to the people whose lives are now saved. For example, with cyclone warnings and evacuations in Bangladesh or Odisha (India), hundreds of thousands of people who might have died in similar cyclones in the past are now 'saved'. But their lives are often ruined, and to avoid these livelihood impacts would require protection and/or replacement of assets and livelihoods (items 5 and 6, in the context of 7) where significant DRR actions are often missing. This is where the emphasis of DRR needs to make a major shift towards what could be termed DRR 2.0. This would be much more disruptive of the systems of power that are beneficial to elites, for example, those who control land and resources in rural Bangladesh and India, and who would resist changes to the allocation of assets and income as part of disaster preparedness. Items 2, 3, and 4 can be (partly) prepared for in DRR approaches but are much more commonly dealt with through emergency relief.

Mapping these different types of hazard impacts onto the seven components of vulnerability above can provide a much more detailed people-centred understanding of what is meant by vulnerability in relation to both a specified hazard (of a particular severity) and what the different types of harm that are possible.

Public health and social causation of disease as a parallel analysis

Some DRR institutions and academics have reverted to the terminology of 'natural' disasters and reduced their emphasis on vulnerability. Paradoxically, in public health there is increasing interest in the analogous processes of the social causation of disease. Looking at this is useful as a metaphor in understanding why vulnerability is a powerful concept and deserves to be restored to a meaningful position in DRR. In public health there is increasing recognition that the likelihood of experiencing ill-health (i.e. being 'vulnerable' to a risk) is mostly a function of similar political, economic, and social processes. 'Natural' factors (a person's biology) are not very significant in determining whether people are healthy or not.

This approach has been championed for more than a decade by the World Health Organization (WHO), which argues that the social determinants of health are the most valid way of understanding differences in wellbeing (WHO 2010). These determinants include income (and its impact on quality of diet, housing, mental stress, education, social networks), housing policy, water and sanitation, the physical environment (including levels of pollution), control of diseases, and whether people can access good health care. These are remarkable in being parallel to the main explanatory factors used in judging vulnerability to natural hazards. A study based on data from the USA considered that the percentage share of health of an individual was only 21% related to genetics and biology, with 79% 'socially determined' (GoInvo, n.d.). The main factors influencing health status related to medical care (access, quality, and availability), the physical environment (including pollution and water quality), and social circumstances (such as access to decent housing, racism, and other forms of discrimination and psychological stress). All of these are directly related to the operation of political and economic factors and the allocation of resources, income, and welfare in a society: tax revenues to pay for health services, water and sanitation, policies on pollution, and housing.

But the remaining category (with a 38% share in determining health status) is related to individual behaviour. Many people harm their own well-being through poor diet, drinking and smoking and using other drugs, lack of exercise, etc. Some also contest what causes illness, and do not accept the 'germ theory of disease'. This was evident in COVID-19 crisis conspiracy theories, and for many people in the West African Ebola crisis of 2013–2016 where there was reluctance to accept that it was a deadly virus. Others both in the global North and South believe in unproven cures and conspiracy theories about vaccination being harmful. Sometimes these beliefs have an element of social causation, involving complex links to systems of power where people mistrust 'big pharma', government, or scientific advice and feel the need to exercise control over their lives by having access to 'popular' alternative explanations. Difficulty in changing behaviours, as for example, with water and sanitation (open defecation, hand washing, food preparation) also arises when there are significant cultural and habitual factors that cause people to be reluctant to change.

These factors are analogous to those involving people with supernatural beliefs about god/s causing disasters, or mistrust of governments (e.g. for exaggerating risks that are low on people's own priorities) in relation to natural hazards. The significance of these behavioural aspects of people's attitudes to the risks of natural hazards is discussed later. This has been a major omission from most standard treatments of the social construction of disaster risk analysis.

Given the adoption of social construction of health in public health policy, it is interesting that for natural hazards the idea of social construction and the importance of the vulnerability concept is now in danger of being diminished. Having been relatively successful in shifting discussion away from the idea that disasters are natural, the recent reversion to 'natural' disasters is interesting and I argue is a consequence of dominant neoliberal policies. By framing problems as a result of 'natural' processes, attention can be diverted away from explaining them in terms of socially constructed vulnerability. They can then be made to appear natural or the fault of the people themselves, as has also happened with explanations of poverty (Gunewardena and Schuller 2008 [which includes several chapters on Katrina]; Vera-Cortes and Macias-Medrano 2020; Giroux 2004).

Who uses 'vulnerability' and why? In what ways is it useful?

For most vulnerability paradigm researchers or organisations, the focus of analysis is on people. The goal of vulnerability analysis is to understand which people (usually in groups sharing particular characteristics related especially to class, ethnicity, and gender) are likely to be the most harmed when a hazard happens. Sometimes users of the vulnerability approach focus instead on different types of entities: infrastructure and buildings are the most common. My preference is to maintain a rigorous people-centred approach that subsumes other affected entities into aspects of people's vulnerability. The 'vulnerability' of a structure is 'converted' in such a way that it shows how when it is damaged the effects are on people and their livelihoods. For example, the possibility that a bridge is damaged (i.e. is 'vulnerable') can be dealt with by examining how it is significant for the people and their systems that use the bridge, especially how damage will affect people's livelihoods.

In the Kobe earthquake of 1995 soil liquefaction caused most of the cranes of the city's container port (sixth largest in the world) to collapse, effectively putting thousands of port workers out of work for many months, along with disruption to tens of thousands of livelihoods dependent on exports (Hamer 1995). People were vulnerable to infrastructure damage because of the impacts on their income. The vulnerability of physical structures is converted to understand how that damage is a source of vulnerability. If people rely on a physical structure's ability to survive a hazard, then part of their vulnerability is incorporated in the potential for damage to the physical structure. A people-centred approach can maintain a clear focus on the harm done to people, their lives, and livelihoods. This can show how entities such as infrastructure and buildings must be analysed in terms of the part they play in livelihood systems. Damage should then not only be measured in the capital loss of the structure (as is the norm), but the loss of the flow of income to those who rely on it in order to live. The vulnerability concept is therefore a key improvement on conventional ways of assessing disaster damage, which otherwise tend to take account only of the capital costs (and may fail to include most poor people's livelihoods which are difficult to cost in terms of dollars).

What makes the term vulnerability valid and useful?

Although it has not been degraded quite so much as words like or 'resilience' (Sehgal 2015) 'sustainability', the term vulnerability is widely used and abused without proper explanation of what it is intended to achieve as a concept (Cannon and Müller-Mahn 2010). It is also used across a range of problems and disciplines (often also without proper specification). Those who work on natural hazards, conflicts, or climate change (as represented in this book) have no right to a monopoly on using the term. Finding ways that the word vulnerability can be used across disciplines and problem areas is probably neither possible nor necessary. This chapter is restricted to assessment of the use of vulnerability in relation to natural hazards, and its possible extension to adaptation to climate change.

The value of the term in relation to hazards and climate change is because it requires an explanation of a disaster (and harm done by climate change) in terms of social causation. It enables a focus on what processes and factors make some people likely to suffer harm when a hazard strikes. There is no standard method for doing this, and below I share my own approach as an attempt to increase the rigour of vulnerability analysis. As mentioned above, a key aspect is to ensure that vulnerability is used in a *predictive* way. Assessment of vulnerability should be made where a hazard is expected to strike, in order to analyse who is likely to suffer harm (and in what degrees and of what type) *before* the hazard strikes. This can be done in a number of ways and scales, ranging from larger-scale risk assessments down to the local ('community') level vulnerability and capacity assessments used by many organisations. The second need is to ensure that the vulnerability is identified in relation to a specific risk: vulnerability must be *hazard-specific* and assessed on the different numbers of people likely to be affected by different levels of severity. A person may be vulnerable to a flood but not an earthquake. They may be vulnerable to a category 5 hurricane, but not a less intense storm.

If these two factors are used in vulnerability analysis it is possible to assess the likely disastrous effects of the specified hazard of a particular severity, so that preventive and preparatory measures can be put in place ('disaster risk reduction'). Otherwise, what is the point of vulnerability analysis? It is not at all useful *after* a hazard has struck to say that its victims must have been 'vulnerable'. By then we know who was vulnerable, and it is too late to prevent the hazard becoming a disaster.

This raises the question as to what term is used for those people who are *not* vulnerable, or whose vulnerability can be *reduced* by preparedness and preventive measures. What term is suitable to describe the 'opposite' of vulnerability? Is it 'resilience', or capacity, or capability? In the 1980s, as NGOs became increasingly involved in DRR, there was an understandable critique of the term vulnerability (exemplified by Anderson & Woodrow in their book *Rising from the ashes* (1989).

They and many NGOs considered that the term vulnerability was too negative, that it emphasised victimhood and did not allow for people's agency and capacities. This was embodied in some of the participatory local assessment methods used by NGOs and the Red Cross/Red Crescent, such as Oxfam's Participatory Vulnerability and Capacity Assessment methodology (PVCA) (Oxfam 2012) and the IFRC Vulnerability and Capacity Assessment (VCA) approach (https://www.ifrc.org/vcaweblinks). I would argue that the concept of vulnerability is akin to the notion of temperature: it is neither about being hot or cold (or vulnerable/not vulnerable), but about a (measurable) scale or spectrum. The problem is that we tend to associate the word itself only with the 'bad' end of the scale.

If people's vulnerability can be identified (in relation to specific types and intensities of hazard) in advance, and disaster risk reduction (DRR) measures put in place, then clearly their vulnerability has been reduced. Does this mean that they have experienced an increase in capacity, or a rise in their resilience? Some of the original users of the term vulnerability did not see it as a negative term that denied agency, but as a scale variable that can allow low and high vulnerability, and any point in between. The point of using the term was to highlight what processes were *causing* vulnerability. The validity of the term lies in its explanatory power. Because it shifts away from nature-based causes and in the direction of *why* people suffer, it can be used to identify what needs to be done with the relevant people to reduce the harm they may suffer. If there is a need to have a term for the opposite of vulnerability, capacity is likely to be the most useful and avoids the problems involved with using resilience.

To achieve its explanatory power, vulnerability needs to be further broken down into sub-components. This is necessary because there are different processes that generate vulnerability, and these operate on different scales and have different qualities. For instance, low levels of income and/or assets (which can be assessed in terms of *class*) are often assumed to produce high vulnerability even though this is not always valid. A different vulnerability factor can counter the advantages that are normally associated with wealth. For instance, upper- and middle-class people were a significant proportion of the victims in the 1999 Izmit (Turkey) earthquake where their apartment buildings collapsed because of poor building quality and failure by construction companies to adhere to building codes (Smith 1999; Schiermeier 1999). By contrast, it would also be possible for poor people to survive in some earthquakes because the collapse of flimsy shacks and houses is less likely to be fatal. In such cases the vulnerability factor is not class and poverty but governance and systems of power that fail to ensure that buildings are safe (probably combined with an attitude that may not take risk seriously). This is significant because the majority of the literature that uses a social causation approach tends to equate vulnerability with poverty (for arguments to the contrary see Wisner et al. 2004). A review of poverty in relation to natural disasters (sic) by authors from the World Bank (Hallegatte et al. 2020) avoids any explanation of disasters (and the thirty years of literature) in relation to social construction, perhaps because it is difficult from their position to comment on the responsibility of governments (and the World Bank) for contributing to vulnerability.

It is therefore unhelpful to think of vulnerability as being a single characteristic of a person, household, or other entity. It needs to be broken down into components, with the additional awareness that a person/household can have less vulnerability on some components and higher vulnerability on others. These differences (when measured through a vulnerability analysis process) can indicate where more effort needs to be made for DRR.

Components of vulnerability

I propose breaking down vulnerability into seven components or major categories.

1. **People's livelihood and available assets**, and the harm that may happen because of a specific hazard – the livelihood's 'resilience' (i.e. what might happen to the livelihood because of a flood of a particular duration, depth, velocity). The amount of income and/or subsistence generated by people's livelihoods is the basis for wellbeing (component 2) and ability to invest in protective measures at the household level (component 3). The amount of income and assets are determined largely by political and economic systems of power (component 5). Each livelihood is also likely to experience hazard impacts in different ways. For instance, a flood may damage land, tools, livestock, and also infrastructure needed to carry out the livelihood. Therefore it is essential to assess how a particular livelihood is likely to be affected by each type of hazard of different intensities and duration.
2. **Baseline status** – the nutritional, health, education and other condition of a person. This is normally determined by the subsistence and/or cash generated by the livelihood, and how well it provides for good scores on these. In some places, the livelihood is supplemented by welfare provided through the government or NGOs, and sometimes by targeted social protection measures. High scores on baseline status are likely to reduce vulnerability to hazards (and climate change) because healthier, well-nourished people (who may also have savings and other assets) are likely to be better prepared and able to recover more quickly.
3. **Self-protection**. This involves the capacity (and *willingness* – see category 6) of people to build a safe home in a safe location in relation to known hazards. Self-protection can lead to reduced risk of death, injury and other impacts (e.g. the ability to store food and water). Self-protection depends on the type of hazard and is not the same for a flood compared with an earthquake. Being able to self-protect is dependent on having the resources to achieve it (for each type of hazard relevant to those people). Those who do not have adequate income and assets are not able to be safe. But a large proportion of people are unlikely to provide for their self-protection even when they have adequate resources. This is because they normally have different priorities or perceptions

of risk (including cultural interpretations) that make them unwilling to spend on safety (see category 6 below).

4. Those who are unable to afford self-protection, or are unwilling to make themselves safe, are therefore entirely reliant on protection through other societal actors in what can be termed **societal protection**. This extra-household protection can take two main forms: the first involves measures that are impossible to achieve at the household (or even local level). These include building codes, warning systems, insurance schemes, and land-use zoning, all of which require national (or even international) initiatives. The second relates to the work of outside agencies (ranging through national or local government to NGOs, Red Cross/Red Crescent, and other civil society organisations) that substitute for people's inability *or unwillingness* to take action to self-protect. Initiatives here could include modifications to houses (e.g. roof straps for storms), raising homes above anticipated flood levels (e.g. earth platforms used in Bangladesh), or group insurance schemes (e.g. crop insurance), and forecast-based cash grants (e.g. the IFRC and Red Cross Climate Centre initiative in several countries) (Costella et al. 2017; IFRC 2014). Different forms of societal protection are highly dependent on the type and effectiveness of the various actors, and this is discussed in terms of the fifth category of vulnerability.
5. This component is more complex and has a double effect and is difficult to name. In the past I have called it **Governance**, using this to signify **systems of power** that are relevant for disasters and vulnerability. The first effect is because systems of power determine the type and amount of societal protection that is possible for a given population and hazard. For example, some governments have not implemented warning systems for hazards even when they are technically possible (as happened in Myanmar/Burma with Cyclone Nargis in 2008) (Gottlieb 2018). In China and some other countries, civil society actors who might work for DRR are discouraged or banned (Human Rights Watch 2019; Ai Weiwei 2018). In others, the implementation of building codes is so corrupted, or rely on inadequate inspection, that the effect of such policies is severely weakened. The second effect of **systems of power** is how they relate to people and their livelihoods. The distribution of assets (such as farmland – in many countries very unequally owned because of the domination of powerful landlords) and income (through redistributive taxation) is often grossly unequal and this leads to poverty through making livelihoods inadequate and welfare provision missing. This means that the first component of vulnerability (adequate assets and livelihood that are protected from hazards and climate change) can be severely constrained by the type of government and other power relations.
6. **Priorities and perceptions**. This relates to how people perceive different types of risks and how they explain extreme events. This is a very complex set of issues that are difficult to summarise, but basically there are two aspects to it. Most people have a very different 'risk hierarchy' than that of the DRR organisations that want to help them to reduce their danger from natural hazards.

For most of the world's people the main problems are daily struggles for food, drinking water, access to health care, and lack of adequate livelihoods (often because of unequally distributed assets and income). The occasional extreme event is not a very high priority (the evidence from thousands of local VCA or similar surveys is that they are given a low priority compared with the problems of everyday life) (IFRC 2014). This is why several authors and organisations consider that it is pointless in developing countries trying to separate DRR from normal problems of development (Wisner et al. 2004; Bradshaw 2014; Collins 2009). This in itself makes it difficult to engage people in DRR activities. Places at risk of severe natural hazards are also often places where people can gain good livelihoods, and from which they would not move even after a major catastrophe (IFRC 2014).

7. Because many people have to 'live with risks', **people's perceptions of disasters** are mostly a result of **cultural interpretations** and often embedded in their religion. Disasters are widely thought of as punishments of god/s, or predestined. Such beliefs about disasters and the related fatalism (while not universal) are therefore a significant part of many people's vulnerability. Priorities combined with beliefs and perceptions of risk are likely to place people in danger (and not just because they are poor and are forced to live in dangerous places), and lack willingness to reduce this. The literature on how culture influences risk perception is increasing, although most DRR organisations tend to ignore culture as a determinant of people's risk behaviour. A word search in eight recent key policy documents of major DRR organisations failed to find the terms *religion/religious, culture, belief* in a total of a thousand pages, suggesting that these factors are not considered relevant to understanding how to reduce disaster risk (the documents are listed after the bibliography). In addition, most DRR initiatives assume that the outsiders' expert intervention to reduce the negative impacts of hazards are what people want (despite a great deal of evidence that this is not the case) (IFRC 2014).

These final two categories of vulnerability deserve further discussion, mainly because they have received much less attention than the others, which regularly feature in work that is based on social construction of disasters.

Livelihood needs, risk perceptions, and the significance of culture

For most work on DRR that has emerged from the vulnerability paradigm, the key issues are forms of social construction of risk based on class, gender, ethnicity, and other power-related social characteristics. It has been difficult to incorporate the last two key factors under 6 and 7, although there is growing academic research on them, including on culture and beliefs about what causes disasters (IFRC 2014; Krüger et al. 2015; Ruehlemann and Jordan 2019).

The first factor relates to people 'voluntarily' living in dangerous places (and houses) and the second relates to how people perceive risk and are influenced by cultural beliefs about disasters. These are of course socially constructed: people's 'willingness' to live with danger can be partly explained as a product of 'constrained choices' and what they believe is possible given their position in society. Supernatural beliefs about what causes a disaster are also embedded in systems of power in which religion itself is a part of the way that elites (and men) manage and control people. Tracing these two factors back to their 'root causes' in systems of power seems crucial but may not provide explanations that are fully traceable to power. Dynes' analysis of the impact of the Lisbon earthquake of 1755 is a fascinating commentary on how using God as an explanation was undermined by scientific understanding and religious doubt, and how this enabled many people to move on from supernatural causes as discussed before (in the shift in some of the world from religious explanations to an understanding of nature as the cause (Dynes 2000).[3] The absence of this shift in much of the world indicates how important it is for DRR to understand this component of vulnerability.

Accepting that they are significant does not need to undermine the social construction of vulnerability paradigm. What is needed is an approach to vulnerability analysis that acknowledges much more the problems of people 'choosing' to live with and in danger, and (in many parts of the world) being enabled to do so by beliefs about the supernatural as a cause of disaster. Underlying this is an awkward reality: many people willingly take risks and place themselves in danger, and do not always invest in greater safety even when they have the resources. In such cases, there is a related problem of potential mistrust of 'the outsiders' who arrive with risk priorities that do not match with the people's own.

This returns us to the analogy of the social construction of illness, where people can demonstrate risky behaviour even when they have access to excellent information about the risks they are taking. As with public health problems, vulnerability analysis needs to take much more account of people's emotional, cultural and behavioural relations with risks. The assumptions of particular forms of 'rationality' in risk-taking in DRR have much to learn from the parallel problems of promoting public health. DRR academics and organisations have not generally favoured assessments of people's own 'need' to live in dangerous places. and have tended to assume that these are constrained choices affected by systems of power involving class, gender, ethnicity, and so on (Cannon 2008). Systems of belief about what causes disasters also have to be much more widely analysed and understood if DRR is to take people seriously on the basis of how they want to deal with risk.

Although both of these 'internal' aspects (people's different priorities, and their belief in supernatural causes of disasters) are actually mostly 'socially determined', the social construction of risk approaches have tended to favour external 'blameable' actors for creating conditions that lead to vulnerability. People placing themselves at risk tended to be overlooked as power systems could not be held responsible so easily. Belief systems, culture, willingness to take risks (trade-offs between hazard and livelihood benefits of a dangerous location) were not much included: the vulnerability paradigm needed someone to 'blame'. Belief systems and willingness to take risks are also socially constructed but not in the same way as they involve an element of 'choice'.

Notes

1 A Google N-gram count in books shows a ten-fold increase in the use of the term *vulnerability* from 1970 to 2018.
2 I use the term 'social factors' here as shorthand for the combination of economic, political, social and behavioural factors – all of which must be seen in relation to the systems of power that generate different levels of likely harm to a particular person or group.
3 In his argument with Voltaire about the disaster, Rousseau was also clear that the event was not 'natural' either: '... nature did not construct twenty thousand houses of six to seven stories [...], and that if inhabitants of this great city were more equally spread out and more lightly lodged, the damage would have been much less and perhaps to no account...' (quoted by Kälin 2011).

References

AAG (Association of American Geographers). 2006. Gilbert White, memorial. http://www.aag.org/cs/membership/tributes_memorials/sz/white_gilbert_f.

Ai Weiwei. 2018. 'The artwork that made me the most dangerous person in China'. *The Guardian* 15 Feb.

Anderson, Mary B. & Woodrow, Peter J. 1989 (1998 reprint). *Rising From The Ashes: Development Strategies In Times Of Disaster.* Boulder, Colorado: Westview Press.

Blaikie, Piers, Cannon, Terry, Davis, Ian, & Wisner, Ben. 1994. *At Risk: Natural Hazards, People's Vulnerability and Disasters.* 1st edition. New York: Routledge.

Bradshaw, S. 2014. *Gender, Development and Disasters.* Cheltenham: Edward Elgar.

Cannon, Terry. 2008. 'Vulnerability, "innocent" disasters and the imperative of cultural understanding'. *Disaster Prevention and Management* vol. 17 No. 3, pp. 350–357.

Cannon, Terry & Müller-Mahn, Detlef. 2010. 'Vulnerability, resilience and development discourses in context of climate change'. *Natural Hazards* vol. 55, pp. 621–635.

Chmutina, Ksenia & von Meding, Jason. 2019. A dilemma of language: "Natural Disasters" in academic literature. *International Journal of Disaster Risk Science* vol. 10, pp. 283–292. doi:10.1007/s13753-019-00232-2.

Christoplos, Ian, L. Tomás Rodríguez, Lisa Schipper, Eddy Alberto Narvaez, Karla Maria Bayres Mejia, Rolando Buitrago, Ligia Gómez, & Francisco J. Pérez. 2009. 'Learning from Recovery after Hurricane Mitch: Experience from Nicaragua – Summary report'. Geneva: IFRC/ProVention Consortium. https://www.recoveryplatform.org/assets/publication/Learning_from_Mitch_summary.pdf.

Cockburn, Patrick. 2011. 'Catastrophe on camera: Why media coverage of natural disasters is flawed'. *The Independent*, 20 January.

Collins, Andrew E. 2009. *Disaster and Development.* New York: Routledge.

Cornwall, Andrea & Eade, Deborah (eds.) 2010. *Deconstructing Development Discourse Buzzwords and Fuzzwords.* Warwickshire, UK: Practical Action Publishing & Oxfam

Costella, Celia, Catalina Jaime, Julie Arrighi, Erin Coughlan de Perez, Pablo Suarez, & Maarten van Aalst. 2017. 'Scalable and sustainable: How to build anticipatory capacity into social protection systems'. *IDS Bulletin* vol. 48, No. 4, pp. 31–46. DOI: 10.19088/1968-2017.151.

DFID (Department for International Development). 2018. 'DFID Response to the Independent Commission for Aid Impact recommendations on: Building Resilience to Natural Disasters', February.

Douglas, Mary & Aaron, Wildavsky. 1983. *Risk and Culture: An Essay on the Selection of Technological and Environmental Dangers*. Berkeley, CA: University of California Press.

Dynes, Russel R. 2000. 'The dialogue between Voltaire and Rosseau on the Lisbon Earthquake: The emergence of a social science view'. *International Journal of Mass Emergencies and Disasters*, vol. 18, No. 1, pp. 97–115.

Giroux, Henry A. 2004. *Terror of Neoliberalism: Authoritarianism and the Eclipse of Democracy*. New York: Routledge.

GoInvo n.d. Determinants of Health, https://www.goinvo.com/vision/determinants-of-health/#references.

Gottlieb, Gregg. 2018. '10 years after, Cyclone Nargis still holds lessons for Myanmar'. *The Conversation* 2 May.

Greenberg, Josh & Scanlon, T. Joe. 2016. 'Old Media, New Media, and the Complex Story of Disasters', in: *Natural Hazard Science*. doi:10.1093/acrefore/9780199389407.013.21.

Gunewardena, Nandini & Schuller, Mark. 2008. *Capitalizing on Catastrophe: Neoliberal Strategies in Disaster Reconstruction*. Walnut Creek, CA: AltaMira Press.

Hallegatte, Stephane, Adrien Vogt-Schilb, Julie Rozenberg, Mook Bangalore, & Chloé Beaudet. 2020. 'From Poverty to Disaster and Back: A Review of the Literature'. *Economics of Disasters and Climate Change* vol. 4, pp. 223–247. doi:10.1007/s41885-020-00060-5

Hamer, Mick. 1995. "Why the Cranes in Kobe Harbour did the Splits." *NewScientist*, https://www.newscientist.com/article/mg14619711-100-why-the-cranes-in-kobe-harbour-did-the-splits/#ixzz7GFY39yl5.

Hartman, Chester & Gregory D. Squires 2006. *There is No Such Thing As a Natural Disaster: Race, Class and Hurricane Katrina*. UK: Routledge.

Hewitt, Kenneth. 1997. *Regions of Risk: A Geographical Introduction to Disasters*. Harlow, UK: Longman.

Horowitz, Andy. 2017 'Don't Repeat the Mistakes of the Katrina Recovery'. *New York Times* 14 September. https://www.nytimes.com/2017/09/14/opinion/hurricane-katrina-irma-harvey.html.

Human Rights Watch 2019 'Human Rights Activism in Post-Tiananmen China - A Tale of Brutal Repression and Extraordinary Resilience'. 30 May. https://www.hrw.org/news/2019/05/30/human-rights-activism-post-tiananmen-china.

ICAI (Independent Commission for Aid Impact). 2018. 'Building resilience to natural disasters'. https://icai.independent.gov.uk/report/resilience/.

IFRC. 2014. *World Disasters Report 2014: Focus on Culture and Risk*. Geneva: International Federation of Red Cross and Red Crescent Societies (IFRC). https://www.ifrc.org/world-disasters-report-2014

Kahneman, Daniel & Tversky, Amos. 1979 'Prospect theory: An analysis of decision under risk'. *Econometrica* vol. 47, No. 2, pp. 263–292. doi:10.2307/1914185.

Kälin, Walter. 2011. *A Human Rights-Based Approach to Building Resilience to Natural Disasters*. Brookings Institute. https://www.brookings.edu/research/a-human-rights-based-approach-to-building-resilience-to-natural-disasters/.

Kates, Robert & Ian Burton. 2008. 'Gilbert F. White, 1911–2006 Local Legacies, National Achievements, and Global Visions'. *Annals of the Association of American Geographers*, vol. 98, No. 2, pp. 479–486. doi:10.1080/00045600801925656.

Kelman, Ilan. 2010. 'Natural Disasters Do Not Exist'. http://www.ilankelman.org/miscellany/NaturalDisasters.doc.

Krüger, Fred, Greg Bankoff, Terry Cannon, Benjamin Orlowski, E. Lisa Schipper (eds.) 2015. *Cultures and Disasters Understanding Cultural Framings in Disaster Risk Reduction*. London: Routledge.

Oliver-Smith, Anthony. 2019. 'Peru's Five-Hundred-Year Earthquake: Vulnerability in Historical Context' in: *The Angry Earth: Disaster in Anthropological Perspective*, Anthony Oliver-Smith & Susannah M Hoffman (eds.) 1999 & 2019 (2nd ed.). New York: Routledge.

Oliver-Smith, Anthony & Susanna Hoffman (eds.) 1999 & 2019 (2nd ed.). *The Angry Earth: Disaster in Anthropological Perspective.* New York: Routledge.

OXFAM 2012. *Participatory Vulnerability and Capacity Assessment: A Practitioner's Guide.* Oxfam. https://oxfamilibrary.openrepository.com/bitstream/handle/10546/232411/ml-participatory-capacity-vulnerability-analysis-practitioners-guide-010612-en.pdf?sequence=4.

Pantti, Mervi. 2018. 'Crisis and Disaster Coverage' in: *The International Encyclopedia of Journalism Studies*. doi:10.1002/9781118841570.iejs0202.

RCCC (IFRC Climate Centre). 2007. *Red Cross/Red Crescent Climate Guide.* https://www.climatecentre.org/downloads/files/RCRC_climateguide.pdf

RCCC (IFRC Climate Centre). 2020. 'Why disaster preparedness cannot wait' https://www.climatecentre.org/news/1298/a-why-disaster-preparedness-cannot-waita

Rich, Nathaniel. 2015. 'Gary Rivlin's *Katrina*: After the Flood'. *New York Times* 5 August. https://www.nytimes.com/2015/08/09/books/review/gary-rivlins-katrina-after-the-flood.html.

Ruehlemann, Anja & Joanne Jordan. 2019. 'Risk perception and culture: implications for vulnerability and adaptation to climate change'. *Disasters* doi:10.1111/disa.12429

Schemper, Lukas. 2019. 'Science Diplomacy and the Making of the United Nations International Decade for Natural Disaster Reduction'. *Diplomatica* vol. 1, p. 2. doi:10.1163/25891774-00102006

Schiermeier, Quirin. 1999. 'Shoddy buildings cost lives in Turkish quake'. *Nature* vol. 400, p. 803.

Secret Development Worker. 2016. 'Secret aid worker: buzzwords are killing development 2016'. *The Guardian* 8 March 2016. https://www.theguardian.com/global-development-professionals-network/2016/mar/08/secret-aid-worker-buzzwords-are-killing-development-human-centred-grassroots.

Sehgal, Parul. 2015. 'The Profound Emptiness of 'Resilience''. *New York Times* 1 December. https://www.nytimes.com/2015/12/06/magazine/the-profound-emptiness-of-resilience.html.

Shaw, Rajib. 2020. 'Thirty Years of Science, Technology, and Academia in Disaster Risk Reduction and Emerging Responsibilities', *International Journal of Disaster Risk Science* vol. 11, pp. 414–425.

Slovic, Paul. 1987. 'Perception of Risk'. *Science* vol. 236, pp. 281–285.

Slovic, Paul & Elke U. Weber. 2002. 'Perception of Risk Posed by Extreme Events', *Conference paper for "Risk Management strategies in an Uncertain World,"* Palisades, New York, April 12–13. https://www.ldeo.columbia.edu/chrr/documents/meetings/roundtable/white_papers/slovic_wp.pdf.

Smith, R. Jeffrey. 1999. 'Turks Blame Inadequate Building Codes'. *Washington Post* 21 August, p. A1.

Telford, John, Margaret Arnold, & Alberto Harth. 2004. 'Learning Lessons from Disaster Recovery: The Case of Honduras'. Hazards Management Unit, Working Paper No. 8, World Bank. https://www.preventionweb.net/files/1595_hondurasWps.pdf.

Varley, Ann (ed.) 1994. *Disasters, Development and Environment.* Chichester, UK: John Wiley & Sons.

Vera-Cortes, Gabriella & Jesus Manuel Macias-Medrano (eds.) 2020. *Disasters and Neoliberalism: Different Expressions of Social Vulnerability*. Springer, https://link.springer.com/content/pdf/bfm%3A978-3-030-54902-2%2F1.pdf.

WHO (World Health Organization). 2010. 'A conceptual framework for action on the social determinants of health'. Social Determinants of Health Discussion Paper 2 (Policy and Practice).

Williams, Nikesha Elise. 2020. 'Katrina battered black New Orleans: Then the recovery did it again'. *Washington Post* 28 August. https://www.washingtonpost.com/outlook/katrina-battered-black-new-orleans-then-the-recovery-did-it-again/2020/08/27/193d2420-e7eb-11ea-bc79-834454439a44_story.html.

Wisner, Ben, Piers Blaikie, Terry Cannon & Ian Davis. 2004. *At Risk: Natural Hazards, People's Vulnerability and Disasters*. 2nd edition. New York: Routledge.

World Bank. 2010. Natural Hazards, UnNatural Disasters: The Economics of Effective Prevention https://openknowledge.worldbank.org/handle/10986/2512.

DRR Policy documents that omit reference to religion, culture, beliefs

Abhas, K. Jha & Zuzana Stanton-Geddes, eds. 2013. '*Strong, Safe, and Resilient: A Strategic Policy Guide for Disaster Risk Management in East Asia and the Pacific*' http://documents.worldbank.org/curated/en/230651468036883533/pdf/758470PUB0EPI0001300PUBDATE02028013.pdf.

Asian Development Bank. 2017. Disaster Risk Assessment for Project Preparation: *A Practical Guide*. https://www.adb.org/documents/disaster-risk-assessment-project-preparation-guide.

Bronkhorst, Van Bernice. 2012. *Disaster Risk Management in South Asia: Regional Overview*. Washington, DC: World Bank. http://documents.worldbank.org/curated/en/648281468170977802/Disaster-risk-management-in-South-Asia-regional-overview.

GFDRR 2017a. *Global Facility for Disaster Reduction and Recovery Strategy 2018–2021*. Washington, DC. https://reliefweb.int/report/world/gfdrr-strategy-2018-2021

GFDRR 2017b. *Work Plan 2018 Bringing Resilience to Scale*. Washington, DC. https://www.gfdrr.org/sites/default/files/publication/2018-gfdrr-work-plan.pdf.

GFDRR 2018 Global Facility for Disaster Reduction and Recovery Annual Report 2017. Washington, DC. https://www.gfdrr.org/sites/default/files/publication/GFDRR-Annual-Report-2017.pdf.

UNISDR 2015. *Making Development Sustainable: The Future of Disaster Risk Management. Global Assessment Report on Disaster. Global Assessment Report on Disaster Risk Reduction*. Geneva, Switzerland: United Nations Office for DRR https://www.unisdr.org/we/inform/publications/42809.

World Bank. 2012. *Disaster Risk Management in Latin America and the Caribbean Region: GFDRR Country Notes*. Washington, DC: World Bank. http://documents.worldbank.org/curated/en/648281468170977802/pdf/763020WP0P11400Box0379791B00PUBLIC0.pdf.

PART II

Vulnerability, conflict, and state–society relations

6

DISASTER STUDIES AND ITS DISCONTENTS

The postcolonial state in hazard and risk creation

Ayesha Siddiqi

Introduction

This chapter draws on two and a half years of research examining disasters in areas affected by conflict and insurgency. The project started in the southern island of Mindanao, in the Philippines, and was later extended to Colombia. The 'lived experience' of disasters in communities living with the everyday physical and structural violence of armed conflict, reveals a cultivated and long-standing relationship between 'natural' hazards and the dominant forces of state and market. This continued experience of marginalisation is at points lived through 'natural' hazards and at different other points experienced as displacement and dispossession, forcing one to re-think the extent to which disasters are a mix of hazards and vulnerabilities and implicating the postcolonial state in the process.

Over the last two decades, much has been written on the relative historical neglect of political questions, within the study of disasters (Olsen 2000, Cohen and Werker 2008; Carrigan 2014 etc.); this has also been accompanied by a lively debate informing a more fundamentally political study of disasters, particularly in their ability to 'transform' political systems in the post-disaster moment (Pelling and Dill 2010; Pelling 2011; Blackburn and Pelling 2018). Much of my own work has been an examination of the latter, through a state–citizen relationship framework trying to better understand how disasters impact social contracts in the postcolonial world (Siddiqi 2013, 2014, 2018, 2019; Siddiqi and Canuday 2018). The argument in this paper emerges from this analytical framework, one that understands risk and vulnerability as defined and 'lived' through the state–citizen relationship. Here, though, I want to draw on Oliver-Smith's (2000) work to challenge the mainstream understanding of processes that make disasters, and use Bonilla's (2020) phrase 'coloniality of disaster' to question, the role of dominant forces

DOI: 10.4324/9781003219453-8

such as the state – especially the state – in not just creating the vulnerability but also the very hazard that drives disasters in the postcolonial world.

Qualitative data for this research were collected at local Barrio (*Barangay*) level in the southern island of Mindanao in the Philippines in 2016–2017. The focus was on one municipality in southeastern Mindanao that was affected by Typhoon Pablo (2012), a Category 5 cyclone, the strongest ever tropical storm to make landfall on the region since records began. This municipality has also been at the centre of the longest-running armed insurgency in Asia, between the Marxist armed group the New People's Army (NPA) and the Filipino state.

In Colombia, data were collected at a town/village level in northern Antioquia Department – in the north west of the country – and Putumayo Department, on Colombia's southern border with Ecuador. Research participants in the towns and villages in the two different regions studied were living with similar forms of conflict dynamics, driven by armed insurgencies such as (now officially demobilised but with dissidents active) *Fuerzas Armadas Revolucionarias de Colombia* (FARC), *Ejército de Liberación Nacional* (ELN), *bandas criminales* (BACRIM), narco-trafficers, and various paramilitary groups, while the hazards that resulted in the disasters were slightly different. In northern Antioquia, research participants had suffered from flooding in May 2018 that displaced tens of thousands of people and left hundreds of other homeless (Daniels 2018). While in the capital city of Putumayo Department, communities had been affected by a deadly landslide that resulted in over 200 casualties and thousands affected.

Re-thinking disasters

Literature tracing the geneology of disaster studies follows a somewhat familiar path. There is usually some mention of the discipline emerging in the latter half of the twentieth century, primarily out of the 'Strategic Bombing Surveys of World War II' (Bolin 2006) and the 'US military's practical concerns about wartime situations' (Quarantelli 1992). Then there is a nod to the growth and development of sociological research in this area that took place primarily against the backdrop of 'Cold War militarization'. Evidence of military thinking and its significant influence in disaster studies can be seen well into the 1970s (Carrigan 2015, 119), from 'contingency' plans taken from war-time scenarios, to the reactive command-and-control style interventions, or even in the extent to which disaster planners were being drawn from the armed forces (Galliard 2019). Finally, it is acknowledged that this historical evolution of the discipline tends to cast a long shadow over the first few decades of work on disasters, which was disproportionately dominated by a focus on studying (i) geophysical processes responsible for hazards, (ii) engineering physical and social 'solutions' to manage these hazardous processes or (iii) emergency and crisis planning to manage the fallout of these processes (Hewitt 1983).

There is also consensus that the critiques of this literature in the 1970s, from a number of contributors to this volume, were instrumental in bringing about a paradigm shift in disaster studies. Carrigan (2015) refers to this scholarship as emerging from 'materialist-inspired geographers, anthropologists and historians such as Kenneth Hewitt, Anthony Oliver-Smith and Ben Wisner' who reframed thinking on disasters, from focusing primarily on the 'natural' *agents* responsible for environmental destruction to the socio-political *processes* putting people in harm's way (120). This vulnerability paradigm moved the conversation away from the physical and extreme events that overwhelm society to the study of power and resources that determine who in society is impacted by disasters and how (O'Keefe, Ken, and Wisner 1976; Hewitt 1983).

Such an agenda was not only a 'critical departure' from the mainstream understanding of disasters but was also radical in the way it placed politics at the front and centre of any study of disasters. Reflecting on the last forty years of disasters research, a recent piece pays tribute to the pioneering works on the vulnerability approach by emphasising their direct relationship to another movement that emerged in the post-war era – postcolonial theory (Fanon 1963; Said 1978). The latter was instrumental in shaping a conceptual understanding of hazard-based disasters, centring on unequal power relations, within societies and especially between the Global North and the postcolonial states of the Global South (Bankoff 2001; Galliard 2019). The vulnerability approach has come to be known as the single most important development that enabled the field of disaster studies to move forward by opening it up to all types of scholarly engagement from outside technical sciences, from literature and cultural studies to economics and political science, hence the importance of this agenda cannot be overstated.

If the study of disasters is a study of complexity, nowhere is this complexity more evident than in trying to grasp the utility of the vulnerability paradigm itself. Recent scholarship has raised important questions on what were once considered the most sacred principles of the study of disasters. In particular, by asking if we have indeed

> grasped the full implications of our critique of the hazard paradigm, or, rather, are we perpetuating some of its core and most problematic tenets? Have we actually taken on the challenge set up for us 40 years ago by the pioneers of the vulnerability paradigm?

Questions that have even led to the worrying conclusion that despite lofty ambitions, radical agendas and the significant influence of postcolonial studies, 'disaster studies still mirrors a Western hegemony that we were meant to contest in the first place' (Galliard 2019).

This recent critical work has tended to particularly highlight the ways in which the 'vulnerability approaches' have been co-opted as systems of knowledge that render 'life amenable to government interventions' (Grove 2014a, 199). These knowledge systems focus on preparing people and societies to manage exposure and discursively construct life as a problem that must be secured against unpredictability. An understanding of vulnerability that assesses degree and scope of exposure to hazards has been hollowed out of its social and political roots (Galliard 2019). Instead, what remains is a 'biopolitical' knowledge system that has been systematically depoliticised (Grove 2013; Galliard 2010), most recently with the emergence of the resilience agenda (Grove 2014a).

The *radical* agenda that the vulnerability paradigm was espousing in its earlier days seems to have been either lost or neoliberalised along the way (Grove 2014b). Thus, in spite of its vision, disaster studies has neither ended up being as radical, nor as critical as had been originally anticipated. In fact, in a departure from its postcolonial theory-inspired roots, disaster studies has even in some instances perpetuated the hegemony of Western thought and ideas in the Global South (Galliard 2019). Even at a time when interventions are increasingly being driven by national authorities, the policy discourse around disaster response is primarily dominated by Western agencies and finances, and 'the postcolonial conception of disaster response as something "we" do on behalf of "them" has proved stubbornly difficult to shift' (Folley 2019).

The critique that disaster studies has busied themselves in setting 'realistic' rather than the 'radical' expectations, set out by the vulnerability agenda, is a valid one. Scholars have defended this position in stating that 'realistic expressions recognise that disasters are too powerful to be prevented, and thus societies need to adapt themselves and their actions to reduce their effects. As a result, climate change adaptation has been used since the mid-1990s' (Davis 2019). In some ways, this thinking echoes the theoretical foundations of the hazards paradigm by accepting that society can only improve its learning on how to live and deal with disasters, rather than challenge the very structural elements that create hazard and risk in the first place. Voices from the Global South have also tended to remain largely absent or under-represented in this area (Folley 2019).

Going further, this chapter illustrates that in complex postcolonial contexts a clear distinction between the *physical* hazard and *social* vulnerability (Cannon 1994) is neither obvious, nor helpful as an analytical tool. This division between a *physical* process that cannot be challenged and a *social* system that can tends to provide a rationale for thinking that disasters are 'too powerful' and overwhelm society. Thus, it shifts collective goals towards 'building resilience' and 'adapting' to hazard-related devastation, rather than addressing the unequal historical processes that result in disasters. This framework that sharply splits a disaster between *natural* hazard and *social* vulnerability has received increased attention in literature tying disasters to anthropogenic climate change. In this context, climate change is viewed as the mother of all natural hazards imbibing a renewed urgency and primacy to hazards thinking and its agenda (Ribot 2014). The problem with this increasingly dichotomised discussion of disasters is that the global is the site where the inevitable and overwhelming hazards take place, while the local is defined almost exclusively by its vulnerability. 'The local is constructed as divergent from the developed North through its vulnerability, defined by its lack of the North's resilience' (Branch 2018, S310).

This dichotomy between a global, natural hazard and local, social vulnerability in the climate change disaster discourse generates a new urgency around disasters that are seen to be linked to impending changing climate and those that are not. Thus, locally significant environmental destruction of varying origins is rendered invisible or less important due to the, perceived far more, urgent threat of global warming-related climate disasters. Even though the vulnerability approach makes clear the need to take the past, the local, and the social into account, the influence of climate-change thinking has inevitably resulted in a disciplinary evolution that privileges this *future*, the *global*, and the *natural* in contemporary disaster studies scholarship (Branch 2018).

Dichotomies – temporal (past versus future), scalar (local versus global), ontological (socially constructed versus natural) have been central to the study of disasters. In fact, newer perspectives have emerged that challenge and critique some of these dichotomies. Increasingly, 'postcolonial disaster studies' (Carrigan 2015), 'disaster colonialism' (Rivera 2021) and 'coloniality of disaster' (Bonilla 2020) literatures seeks to resolve the tensions between (environmental) event and (historical) process by situating the former firmly within the wider processes of exploitation and subjugation. Yet it has also been noted that dichotomies present 'the greatest obstacle' to a nuanced understanding of hazards related disaster today (Branch 2018). The sharp analytical breaks between social and natural, local and global, and past and future results in disaster thinking seeking to address those parts of the equation that seem most amenable to technical interventions, rather than engaging with a more 'multi-dimensional, context specific and historically relevant understanding of climatic disasters' (Siddiqi 2014). Drawing on this critique of dichotomised disaster analysis, I want to suggest that hazards and vulnerability are better understood as interwoven, deeply overlapping phenomena subject to similar constitutive social processes that emerge from the very nature of the social contract between the postcolonial state and its (subaltern) citizens.

In the 'complex' and 'compound' disasters taking place in the postcolonial world, hazards are often far less *natural* and can only be adequately understood within the context of long-existing racial and colonial inequalities (Bonilla 2020), formalised by the state–citizen social contract in the postcolonial world. Recognising the politics of *hazard creation* enables disaster studies to address two of its most primary critiques. Firstly it enables a better representation of the concerns, if not the voice, of the subaltern who does not speak of the floods and typhoons as singular and isolated events but rather as part of a longer history of oppression and marginalisation by state and market forces – the disaster of colonialism. And secondly, it provides another analytical toolbox for moving attention away from the inevitable physical and natural hazard that overwhelms state and society.

My own work in areas where people have lived experience of disasters amidst conflict and violence illustrates particularly powerful cases of why and how there is a need to go beyond these dichotomies and towards the oppressive powers of state forces. In fact, this relationship between political violence and climate disaster highlights the need to rethink the very concept of disaster as an single event and towards what scholars have called 'devastation' (Branch 2018), 'catastrophe' (Carrigan 2015), 'slow violence' (Bonilla 2020) and 'structural violence'. A reading of 'multi-scalar vulnerabilities' and 'multiple histories' (Mika 2019) of hazards that culminate in deliberate and orchestrated forms of violence through disasters. The remaining chapter will present two empirical case studies based on fieldwork done in the Philippines and Colombia, situating the postcolonial citizen's relationship with its state at the centre of the politics of hazard creation in these local muncipalities.

Methods

My research in the Philippines was primarily ethnographic. I did fieldwork with local researchers and NGO partners in a *barangay* devastated by Typhoon Pablo (2012), over a period of five months between December 2016 and September 2017. This was also complemented by digital storytelling workshops in two neighbourhoods. Interviews were also used to collect perspectives of policy makers and practitioners working on disasters and the conflict with the NPA, at the municipal, provincial, and national levels.

In Colombia, interviews with policy makers were conducted primarily in Bogota and Mocoa, while interviews with community members took place in Mocoa and four towns in Northern Antioquia (Sabanalarga, Toledo, Ituango, and Puerto Valdivia) affected by flooding in 2018, due to dam construction. Undertaking fieldwork in the latter is particularly complicated for non-Colombians due to accessibility and security issues and the dangers of political and state violence have to be 'managed' (Sluka 2012, 2015). It was therefore essential, for the purposes of speaking to affected people and constructing their narrative, to accept the invitation of a local civil society organisation, *Rios Vivos* (Living Rivers), taking international observers to the most flood-affected towns.

This method of engaging with flood-affected residents in northern Antioquia had some obvious limitations. As researchers, we were primarily able to speak to residents connected to this activist civil society organisation, pursuing a very particular anti-dam political agenda. In all four localities, these were people who had come to participate in the town hall meetings organised by *Rios Vivos*, we were able to document the testimonies of these residents and follow-up with more individual interviews after the meetings. At the same time recognising the impossibility of undertaking this research independently and as 'outsiders' in communities where people receive frequent and regular death threats, a calculated decision and 'pragmatic strategy' (Kovats-Bernat 2008) was made to use the help of *Rios Vivos*, while at the same time being reflective, to whatever extent possible, of its impact on data collection.

The state and the politics of hazard creation in the Philippines

In the aftermath of Typhoon Pablo in 2012, the municipal state in one province of southeastern Mindanao relocated an entire village community of disaster-affected people to a high gradient 'plateau' citing safety concerns. Forest cover was quickly removed from the 'plateau', and close to 100 brick and mortar houses were constructed. The state provided cash grants to purchase building materials necessary for the construction of these homes and labour was provided free of charge as goodwill by family and friends. While doing interviews in this community it became evident that the municipal state – from local council representatives, municipal DRR officers, the mayor, and department of social and welfare – was pushing the rapid development of this residential housing community on the 'plateau'. The public narrative widely encountered in this area was that this state-level push for re-location was based on the results of scientific assessments carried out by the national hazard mapping agency, the Mines and Geological Bureau (MGB). These assessments had classified the original, pre-disaster location of the neighbourhood, devastated by Typhoon Pablo, as 'extremely hazardous' and deemed unsafe for human settlement.

An issue that was never raised in the official state narrative was that resettling these households away from their typhoon-affected homes was also clearly linked to another political agenda. A 'hydro company' had moved into the area, a year after the typhoon, to construct a dam on a part of this river and a new military checkpoint had been established in the abandoned village to protect the equipment and investment of the company, Euro Hydro Power. Residents now living on the 'plateau' sometimes implied that their resettlement away from their old village had more to do with protecting corporate interests rather than ensuring their safety from hazards. As one elderly woman in the resettled community said:

> The military checkpoint has not been based near our abandoned home for us. It is not us that they are guarding its the hydro company. They are the protector of the company, Euro Hydro. Moving everyone to this 'plateau' will make it easy for them to protect that area (from insurgents).

Over the time that the research team spent during fieldwork in this region other motives for re-settling an entire residential community away from the original village started becoming clearer. This move to a high gradient forested 'plateau' seemed to not be driven entirely by 'risk reduction' motives. The state also needed to open up the hydroelectricity market for the company Euro Hydro Power, Asia Holdings. Interviews with the local municipality revealed how ill-advised and hazardous the re-settlement on the 'plateau' had been. The Municipal Disaster Risk Management (MDRM) Officer explained that this plateau was in fact highly vulnerable to landslides.

> One of the problems we now face, and it is a challenge to our office, because we have recently assessed the area, the 'plateau' is a landslide prone area, considering the slope and mountainous region...

There was also some recognition amongst staff at this local level that removing natural vegetation and forest cover to construct new buildings has definitively increased hazard-risk in this area. At the same time, local residents were also expressing their own feelings of living with increased risk. In fact, some individuals were clear, that in planning such a residential development on this plateau, one that resulted in deforestation and increased inundation every time it rained, new forms of hazards were being actively being created by the state. A disabled middle-aged resident from the area, who walked with the help of crutches, was particularly concerned about the way their re-settlement had placed them in the path of a new kind of 'natural' hazard and would also result in creating this hazard to begin with.

> I worry about our house because we live on the gradient of the hillside and I worry this (construction) will inevitably cause a landslide. I also observed that the culverts at the bottom of the hill 'overflow' during heavy rains. I know these are small worries, but I still worry about them. I also know these are just my speculations. That is why when there are signs of flooding, we get very scared, considering we live on a steep hill. Since there are no trees here anymore, the ground also gets water-logged and turns muddy so we have to stay on guard and be ready (to evacuate) when it rains hard.

Research suggests that 'heavy rainfall transform(s) simple local floods and debris flows to large-scale landslides and "compound disasters"' (Chen et al. 2011, 1). The term 'compound disasters' refers to the process whereby multiple environmental drivers, such as small-scale floods, debris flows, landslides, and landslide lakes converge, creating a single disaster. This research also suggests that in case of typhoons in the Philippines this is not only a common occurrence but also one that is increasing in frequency and intensity due to human interventions (Chen et al. 2011). It is almost certain that in the coming months, or years the 'natural' hazard of heavy rains, will converge with all these other environmental variables to create a compound disaster on the 'plateau'. Most people on the field site, from officials working at the MRDM office to residents living on the 'plateau', are acutely aware it is going to happen and are expecting its devastation. There is little doubt that this hazard too will be part of the longer story of state marginalisation and dispossession of the indigenous peoples in Mindanao.

Infrastructure and the politics of risk production in Colombia

The floods in and around Ituango were the result of structural failures in Colombia's largest Hydroelectric dam project under construction on the Cauca River. The dam was being built by *Empresas Públicas de Medellín* (EPM), a public utility corporation owned by the City of Medellin in Colombia. Historically Bajo Cauca also has a high proportion of coca cultivations, making it a focal point of armed conflict, which has continued in the post-Accord years. The years 1996 and 1997 were particularly bloody and painful. During this time the paramilitaries committed massacres against the civilian population, including in El Aro and La Granja (in Ituango municipality), claiming they were guerrilla sympathisers

(Rutas del Conflicto 2018). According to a well-known disaster risk reduction specialist in Colombia, the perceived link between the massacres that took place about 10 years before the dam project started and the environmental disaster as a consequence of this anthropogenic intervention on the river, makes this Colombia's most complex disaster ever. In fact, he somewhat cynically remarked that 'whenever there is a massacre, we know that a mega (development) project is coming', implying that the violence is directly perpetrated by state-led forces pushing this vision of development.

Primary research included testimonies gathered in different 'town hall' style meetings in four particularly badly affected communities, followed up by individual interviews. These conversations revealed that the history of armed conflict and the displacement and dispossession due to the flooding disaster is one unbroken continuum of violence in people's narratives. The hazard of heavy rains (and dam malfunction) is not separate, unrelated or an additional event that happened, apart from all the others relating to the conflict and violence in the region. It is part of the same process that includes wider denial of rights and marginalisation by state and market forces. People in northern Antioquia and Bajo Cauca feel they are subjected to this oppressive experience of citizenship for being *campesinos*, considered 'FARC sympathisers' and marginal to the interests of the political elite in Bogota. Residents in this part of the country are no strangers to state brutalisation and violence, historically acting through paramilitary forces during the dark years, at the peak of the armed conflict in the 1990s. The degree to which local citizens hold the state accountable for the extreme violence of yesteryears varies area to area and household to household but at the very least it is believed that the state is complicit in looking the other way when these atrocities were being committed. During fieldwork in these towns we encountered numerous individuals who had survived, or fled massacres in the late 1990s, early 2000s in Northern Antioquia. Their narrative of the floods was only one part of their wider story.

As one man in Puerto Valdivia, the town most evidently and visibly destroyed by the flooding of May 2018, said 'we have lived with massacres our whole lives'. In his view, historically, the paramilitaries came, committed their atrocities on local residents thought to be siding with guerrillas and then moved on. Locally, he believed people had learnt to live with that 'normal'. What was far less usual or normal for him, was the scale of the displacement experienced during these floods in May 2018. Branch's (2018) work on Northern Uganda poignantly captures the various ways in which the experience of a climate-related disaster, in a region with a long and tortured history of violent insurgency and equally brutal state counterinsurgency, dissolved the line between a 'global natural hazard' and 'local social vulnerability'. He refers to the drought in Northern Uganda as embedded within the legacies of war – 'a form of violence that continued into the post-conflict period'. There is little doubt that the narrative around the flooding of River Cauca that our interviewees were constructing – with the help of an activist movement such as *Rios Vivos* – was one where this disaster was a continuation of state violence, through other means, in the postcolony.

In fact, one middle-aged man in Ituango said that despite being led to believe that the peace deal signed with FARC in 2016 would lead to an improved security and a better life, he had only seen things getting worse. This was because a large number of environmental defenders and leaders of their movements were either disappearing or winding up dead. In all testimonies and interviews it became evident that those addressing the research team and international observers wanted to express how deeply painful and difficult the flooding disaster – and going back further the construction of this dam – had been in the face of a state and EPM narrative that was 'de-catastrophising' the disaster (Warner 2013). At the same time, interviewees were making clear the fact that this was not an isolated or unusual case of marginalisation experienced by them at the hands of the state.

In most cases, interviewees were clear in their implication that the state is an enabling agent for EPM (a public corporation owned by the city of Medellin), much like for decades it managed to enable paramilitaries to fight the war against the guerrilla groups. As Branch (2018) points out then,

> it is not the state's response, or lack thereof, to climate disaster that has political meaning; rather, the drought itself is seen as political, as a continuation of state violence. By being part of longer processes of social–natural devastation, climate disaster may not comprise a disruption, but rather an intensification of, and continuity with, these existing political arrangements and structures.

In the town hall meeting in Toledo, one woman said that even when people began to arrive at town meetings in the early days after the floods, they were unable to find their words to speak. After being terrorised by the conflict for so long, they had no idea how to write petitions and engage with the state. In a particularly moving moment in the meeting in Puerto Valdivia an older man in his sixties and a survivor of violence and torture at the hands of paramilitary forces, said it was so traumatic to see his whole town washed away by floodwaters he had to seek professional help from a psychologist. The fact that these devastating floods were another chapter in the wider story of state marginalisation and oppression of those who consider themselves 'campesinos' was never far from the surface of these conversations.

> People's desperation and insecurity in the face of state-violence-driven climate disaster thus does not lead to conflicts among rural people, but to a confirmation of their oppression by the state. One should not view potential future violence as resource conflict in response to climate disaster; instead, it should be seen as a continuation of the violence of the war, perhaps even involving new armed opposition to the state.
>
> (Branch, 2018)

In some interviews the anger towards the state, particularly opposition to the sorts of development paradigms it was following and the ways in which it was evolving the methods through which to perpetuate violence against these citizens, was palpable. One middle-aged woman, who had been relocated to a local gymnasium with other families affected by flooding, said that she felt trapped in her makeshift home by the state. Over the last 25 years she had lost three family members to the conflict. Her husband had disappeared and was presumed killed after which two of her brothers had been tortured and killed by unknown forces, believed to be paramilitaries. She was emphatic in stating that the armed conflict, in this region, was linked to the flooding of May 2018, through the dam. In her interview, she stated that the paramilitaries killed all the young men who could present a challenge and then 'the company' (EPM) came in. 'The state helped the company – you can't see it any other way. They have killed many, many men.' In this narrative, it was repeatedly stated that the state uses legality and markets to justify oppression. As the case of Bajo Cauca makes evident disasters are not just the result of vulnerability but of multiple and compounding hazards created by imperial state and market forces.

Competing visions of risk reduction and risk creation between the state and its citizens

In both field sites in the Philippines and Colombia relocating people away from 'high-risk' hazard areas was a regular policy pursued by the state. Conversations with government officials and those at the helm of policymaking also regularly referred to re-settling and re-locating people, away from high-risk areas to reduce their vulnerability to 'natural' hazards. This has been a fairly standard state policy, especially in informal settlements as noted by other scholars (Fraser 2017; Zeiderman 2016). In these interviews, the fact that the people living with disasters, saw this as a much longer process of victimisation and brutalisation of the state was never a subject of official reflection or introspection.

In fact, there seemed to be a single-minded determination on the part of state officials to move forward with technical 'planning' and implementation of official 'risk reduction' in disaster-affected localities in both field sites. On the ground, this agenda seems to be pushed along even at the expense of those whom it is meant to protect. While an official from Putumayo was confident of his well-resourced agency and the technical 'knowledge' they had available, he suggested that it was in fact people who were the problem.

> More than that, every day, more people locate themselves in zones of risk… if the community doesn't take a role in risk management, the risks will continue. It is hard to kick people out of the high-risk zones. People can be dangerous as well.

When speaking of how the re-settlement land, where the community affected by the landslide in Mocoa was re-located to, was acquired:

> The UNGRD[1] bought it. The municipality is autonomous in saying where, but people need the UNGRD because it has the resources to buy it. There are various mechanisms. 1) you sell it to me. 2) If you don't want to sell it, I'll expropriate it under a judicial sentence. When it is of public interest and if you don't want to sell it, I'll expropriate it.

A hard engineering-driven disaster risk reduction paradigm that over-emphasised re-location and re-settlement at the expense of other strategies was often the clearest message coming from state authorities. Seemingly, partly as a result of high-modernist state policies 'best conceived as a strong, one might even say muscle-bound, version of the beliefs in scientific and technical progress' (Scott 1998), and partly a legacy of decades of conflict and millions of displaced and re-settled people (Zeiderman 2012).

Beyond simply differing perspectives on vulnerability between the postcolonial citizen and the state, Foucaudian ideas on the politics of life suggest something quite different. The imperative to 'make life live or let die' are directly linked to sustaining life of value to the market (Anderson 2011). This work enables an understanding of the philosophical foundations for how hazards much like war are needed to ensure security – killing life to protect *valued* life. Not all life but *productive* life must be secured from unexpected and unanticipated disruptions to productive activity. Yet, 'it is not that life has been totally integrated into techniques that govern and administer it; it constantly escapes them' (Foucault 2008, 143 in Anderson 2011) and the state – as the centre of political control -must constantly evolve and mutate for the successful 'subsumption of life'. In other words, re-location and re-settlement are not just tools for governing emergencies but instead by ensuring some people's lives are moved ensures other *valued* lives are protected and cared for (Anderson 2011).

A media analysis of five regional and national newspapers[2] in the aftermath of Typhoon Pablo reveals over twenty news stories indicating that the state was moving people from the highlands and affected *barangays* to 'evacuation centres' and 'relocation sites.' These new sites were always reported to be safely away from Davao – the capital of Mindanao and most other urban centres. Placing these indigenous and marginal communities in 'structurally strong and environmentally friendly shelter units which can withstand wind velocities up to 180kph, earthquakes up to intensity four…' (SunStar Davao, 4 February 2013) in previously forest-covered and non-habitable areas is likely putting them on the frontlines of new hazards while ensuring *valued* life in Davao is secure.

In historically 'periphery' regions of Colombia and Philippines the ideational and ideological conflict over hazards created by state and market forces, what is often articulated as a brutal state or 'company' destroying local ecology, can be understood as dominant forces protecting, caring for and sustaining valued lives by abandoning, damaging and destroying other lives (Anderson 2011).

Nowhere was this more evident than in the town hall meeting in Toledo, a town in northern Antioquia, destroyed by flooding in 2018. Every single resident who spoke about their experience of the disaster, first and foremost mentioned the devastating impact on local ecology. There were regular references to parrots whose habitat has been destroyed by flooding and honeybees that disappeared overnight because trees were mercilessly cut for dam infrastructure. Residents were distraught that coca production in the region, grown by armed groups to fund the conflict, had already resulted in significant deforestation making them more vulnerable to the flooding. Similarly, in the Philippines, interviewees who had been relocated to the new settlement repeatedly spoke of a lack of understanding at the state level of local habitat and vegetation that had been bulldozed to the ground to create this new residential community. For a number of respondents living without any natural vegetation in a tropical climate in a naturally forested area was a deeply traumatic experience forced upon them by the state.

In these countries where there is a history of state violence against citizens this ideational conflict evidently put the state–citizen relationship under further stress. In fact, these incompatible visions of DRR were often experienced by people in the form of an 'uncaring', even 'brutal' state, whose policies result in dispossession and displacement and result in creating 'natural' hazards that continue the victimisation of citizens. In both Colombia and Philippines, the narrative around a state that is not concerned in the well-being of the *campesinos* (peasants) or the people of the *bukit* (highlands) respectively, was prevalent.

Conclusion

This chapter has argued for complicating the understanding of hazards as 'natural', demonstrating through empirical data that on the margins of the postcolonial state, the physical events (of the landslides or floods) are created by dominant state and market forces, based on life that is valued and that which it can 'let die'. It has illustrated this, using the case of contentious re-settlement in southern Philippines likely to result in landslides and infrastructural interventions in northwestern Colombia that resulted in large-scale floods. Hazards that cannot simply be attributed to natural forces but rather to political choice made by an active and engaged state. Citizens on the peripheries of the postcolonial state (and of marginal relevance to the market) have highlighted that these hazards are only the most recent form of violent disaster they have experienced in the history of their relationship with the state.

In an interview with the head of the *Rios Vivos* movement we asked her, 'do you think if the dam was built somewhere else in the country, the displacement and dispossession would have...' Even before we could finish the question she replied, 'but that's our point, make no mistake, somewhere else in the country – this dam would never have been built.' In small towns and villages of the postcolonial state, away from the high rises of Medillin or Manila, communities self-identify as *dispensible* – lives that are chosen to be placed in harm's way – almost as much as they identify as indigenous peoples, or belonging to a particular ethnic group.

Recognising that creating hazards and risk are active tools of governmentality to manage subaltern populations enables a deconstruction of the politics of hazard creation. In so doing they provide the tools and the language with which to address some of the shortcomings of contemporary disaster studies. Citizens on the margins of the state experience the coming together of familiar processes of exclusion, denial of rights, and disempowerment in these floods and landslides. Subaltern voices in Philippines and Colombia discursively construct hazards as a political choice, illustrating their own dispensability to the postcolonial state and market forces continuing their long-standing marginalisation by these dominant actors. For disaster studies to reconcile with postcolonial subjects, on the margins of the state, it must open itself up to ontologies that see hazard creation as less physical and more part of a longer history of oppression in the postcolonial world. Thus, in this way, it takes away some of its power as overwhelming physical processes.

This chapter therefore demonstrates how hazards and risks can be understood as deeply rooted in the state–citizen social contract, part of long-standing social processes of exclusion and marginalisation in the postcolony.

Notes

1 *Unidad Nacional para la Gestión del Riesgo de Desastres* (National Unit for Disaster Risk Management).

2 SunStar Davao, Davao Today, Mindanao Times, Philippine Daily Inquirer, the Philippine Star over three months from 5 December 2012.

References

Anderson, Ben. 2011. "Population and affective perception: Biopolitics and anticipatory action in US counterinsurgency doctrine". *Antipode* 43, Issue 2: 205–236. doi:10.1111/j.1467-8330.2010.00804.x.

Bankoff, Greg. 2001. "Rendering the world unsafe: "Vulnerability" as Western discourse". *Disasters* 25, Issue 1: 19–35. doi:10.1111/1467-7717.00159.

Blackburn, Sophie and Mark Pelling. 2018. "The political impacts of adaptation actions: Social contracts, a research agenda". *WIREs Climate Change* 9, Issue 6: 1–8. doi:10.1002/wcc.549.

Bolin, Bob. 2006. "Race, class, ethnicity, and disaster vulnerability". In *Handbook of Disaster Research*, edited by Rodriguez, Havidan, Enrico Quarantelli and Russell Dynes 113–129. New York: Springer.

Bonilla, Yarimar. 2020. "The coloniality of disaster: Race, empire, and the temporal logics of emergency in Puerto Rico, USA". *Political Geography* 78 (1–12). doi:10.1016/j.polgeo.2020.102181.

Branch, Adam. 2018. "From disaster to devastation: Drought as war in northern Uganda". *Disasters* 42, Issue S2: S306–S327. doi:10.1111/disa.12303.
Carrigan, Anthony. 2014. "Catastrophe and Environment". *Special issue of Moving Worlds: A Journal of Transcultural Writings* 14, Issue 2.
Carrigan, Anthony. 2015. "Towards a postcolonial disaster studies". In *Global Ecologies and the Environmental Humanities: Postcolonial Approaches*, edited by Elizabeth DeLoughery, Jill Didur and Anthony Carrigan 117–139. Abingdon: Routledge.
Chen, Yu-Shiu, Yu-Shu Kuo, Wen-Chi Lai, Yuan-Jung Tsai, Shin-Ping Lee, Kun-Ting Chen, and Chjeng-Lun Shieh. 2011. "Reflection of Typhoon Morakot – the challenge of compound disaster simulation". *Journal of Mountain Science* 8, Issue 4: 571–581. doi:10.1007/s11629-011-2132-5.
Cohen, Charles and Eric D. Werker. 2008. "The political economy of 'natural' disasters". *Journal of Conflict Resolution* 52, Issue 6: 795–819. doi:10.1177/0022002708322157.
Cannon, Terry. 1994. "Vulnerability analysis and the explanation of 'natural' disasters". In *Disasters, Development and Environment*, edited by A. Varley. New York: Belhaven.
Daniels, Joe Parkin. May 16, 2018. "Colombia: Tens of thousands ordered to evacuate after floods at dam". *The Guardian Online*. https://www.theguardian.com/world/2018/may/16/colombia-tens-of-thousands-of-ordered-to-evacuate-after-floods-at-dam.
Davis, Ian. 2019. "Reflections on 40 years of *Disasters*, 1977–2017". *Disaster* 43, Issue S1: S61–S82. doi:10.1111/disa.12328.
Fanon, Frantz. 1963. *The Wretched of the Earth*. New York: Grove Press.
Folley, Matthew. 2019. "Editorial introduction: 40th anniversary special issue of *Disasters*". *Disasters* 43, Issue S1: S3–S4. https://onlinelibrary.wiley.com/doi/10.1111/disa.12329
Foucault, Michel. 2008. *The Birth of Biopolitics: Lectures at the Collège de France, 1978–1979*. London: Palgrave Macmillan.
Fraser, A. 2017. "The missing politics of urban vulnerability: The state and the co-production of climate risk". *Environment and Planning* 49(12): 1–18.
Galliard, Jean-Christophe. 2010. "Vulnerability, capacity and resilience: Perspectives for climate and development policy". *Journal of International Development* 22, Issue 2: 218–232 doi:10.1002/jid.1675.
Galliard, Jean-Christophe. 2019. "Disaster studies inside out". *Disasters* 43, Issue S1: S7–S17 doi:10.1111/disa.12323.
Grove, Kevin J. 2013. "From emergency management to managing emergence: A genealogy of disaster management in Jamaica". *Annals of the Association of American Geographers* 103, Issue 3: 570–588. doi:10.1080/00045608.2012.740357.
Grove, Kevin J. 2014a. "Adaptation machines and the parasitic politics of life in Jamaican disaster resilience". *Antipode* 46, Issue 3: 611–628. doi:10.1111/anti.12066.
Grove, Kevin J. 2014b. "Biopolitics and adaptation: Governing social and ecological contingency through climate change and disaster studies". *Geography Compass* 8, Issue 3: 198–210. doi:10.1111/gec3.12118.
Hewitt, Kenneth. 1983. "The idea of calamity in a technocratic age". In *Interpretations of Calamity from the Perspective of Human Ecology*, edited by Kenneth Hewitt 3–30. London: Allen & Unwin.
Kovats-Bernat, Christopher J. 2008. "Negotiating dangerous fields: Pragmatic startegies for fieldwork amid violence and terror". *American Anthropologist* 104, Issue 1: 208–222. doi:10.1525/aa.2002.104.1.208.
Mika, Kasia. 2019. *Disasters, Vulnerability and Narratives: Writing Haiti's Futures*. Abington, Routledge.
O'Keefe, Phil. Ken, Westgate, and Wisner, Ben. 1976. "Taking the naturalness out of natural disasters". *Nature* 260: 566–567.

Oliver-Smith, Anthony. 2000. "Peru's 500 year earthquake: Vulnerability in historical context." *The Angry Earth*, edited by A. Oliver-Smith and S Hoffman. New York: Routledge.

Olsen, Richard S. 2000. "Towards a politics of disaster: Losses, values, Agenda, and Blame". *International Journal of Mass Emergencies and Disasters* 18, Issue 2: 265–287.

Pelling, Mark. 2011. *Adaptation to Climate Change: From Resilience to Transformation.* Abingdon: Routledge.

Pelling, Mark and Kathleen Dill. 2010. "Disaster politics: Tipping points for change in the adaptation of sociopolitical regimes". *Progress in Human Geography* 34, Issue 1: 21–37. doi:10.1177/0309132509105004.

Quarantelli, Enrico. 1992. "The importance of thining of disasters as social phenomena". *Disaster Research Paper Preliminary Paper 184.* Newark, University of Delaware.

Ribot, Jesse. 2014. "Cause and response: Vulnerability and climate in the Anthropocene". *The Journal of Peasant Studies* 41, Issue 5: 667–705. doi:10.1080/03066150.2014.894911.

Rivera, Daniella Zoe. 2021. "Disaster colonialism: A commentary on disasters beyond singular events to structural violence". *International Journal of Urban and Regional Research* doi:10.1111/1468-2427.12950.

Rutas del Conflicto. 2018. "Masacre de El Aro." *Rutas Del Conflicto.* http://rutasdelconflicto.com/interna.php?masacre=25.

Said, Edward W. 1978. *Orientalism.* New York: Pantheon Books.

Scott, James C. 1998. *Seeing Like a State: How Certain Schemes to Improve the Human Condition have Failed.* New Haven, London: Yale University Press.

Siddiqi, Ayesha. 2013. "The emerging social contract: state–citizen interaction after the floods of 2010 and 2011 in southern Sindh, Pakistan". *IDS Bulletin* 44, Issue 3: 94–102. doi:10.1111/1759-5436.12036.

Siddiqi, Ayesha. 2014. "Climatic disasters and radical politics in southern Pakistan: The non-linear connection". *Geopolitics* 19, Issue 4: 885–910. doi:10.1080/14650045.2014.920328.

Siddiqi, Ayesha. 2018. "'Disaster citizenship': An emerging framework for understanding the depth of digital citizenship in Pakistan". *Contemporary South Asia* 26, Issue 2: 157–174. doi:10.1080/09584935.2017.1407294.

Siddiqi, Ayesha. 2019. *In the Wake of Disaster: Islamists, the State and a Social Contract in Pakistan.* Cambridge: University of Cambridge Press.

Siddiqi, Ayesha and Jowel P. Canuday. 2018. "Stories from the frontlines: Decolonising social contracts for disasters". *Disasters* 42, Issue S2: S215–S238. doi:10.1111/disa.12308.

Sluka, Jeffrey A. 2012. "Reflections on managing dangerous fieldwork: Dangerous anthropology in belfast". In *Ethnographic Fieldwork: An Anthropological Reader*, edited by Carolyn Nordstrom and Antonius Robben 283–297. Berkley: University of California.

Sluka, Jeffrey A. 2015. "Managing danger in fieldwork with perpetrators of political violence and state terror". *Conflict and Society* 1, Issue 1: 109–124. doi:10.3167/arcs.2015.010109.

Warner, Jeroen. 2013. "The politics of catastrophisation". In *Disaster, Conflict and Society in Crisis: Everyday Politics of Crisis Response*, edited by Dorothea Hilhorst 76–94. Abingdon: Routledge.

Zeiderman, Austin. 2012. "On Shaky ground: The making of risk in Bogotá." *Environment and Planning A* 44, Issue 7: 1570–1588 doi:10.1068/a44283.

Zeiderman, Austin. 2016. "Prognosis past: The temporal politics of disaster in Colombia." *Journal of the Royal Anthropological Institute* 22, Issue S1: 163–180. doi:10.1111/1467-9655.12399.

7

HUMANITARIANISM

Navigating between resilience and vulnerability[1]

Dorothea Hilhorst

Humanitarianism and vulnerability

Vulnerability has always been a key notion for humanitarian action. One of the fundamental humanitarian principles concerns impartiality and dictates that aid must be based on needs alone. Vulnerability analysis is the way to establish needs. International humanitarian law also brings about entitlements to humanitarian services, in particular the refugee convention and the soft humanitarian law of the Internally Displaced Persons principles. This fundamental embrace of vulnerability has been eroded and obscured in the last few decades because of a change in approach towards resilience humanitarianism (Hilhorst 2018).

From classical to resilience humanitarianism

Humanitarian aid has long been dominated by a paradigm rooted in exceptionalism, grounded in the ethics of humanitarian principles, and centred on United Nations (UN) agencies and international non-governmental organisations (INGO). In recent years this 'classical Dunantist paradigm' has been paralleled and partly overtaken by a radically different paradigm, which can be called the 'resilience paradigm'. Whereas the classical paradigm centres on principled aid, the resilience paradigm foregrounds building on local response capacities. Both paradigms have a strong logic that dictates a specific way of seeing the nature of crisis, the scope of the humanitarian response, the identity of humanitarian actors, and the nature of institutions and people in crisis-affected areas. They result in different bodies of practice that can be labelled 'classical humanitarianism' and 'resilience humanitarianism' (Hilhorst 2018).

DOI: 10.4324/9781003219453-9

The classical paradigm of Dunantist humanitarianism has dominated conversations among humanitarians for almost 150 years, despite contestation from concerned scholars and from within the domain (such as do no harm; listening projects; linking relief to rehabilitation and development; and the rights-based approaches that gained popularity in the 1990s but were largely silenced when the 'war on terror' began). For some years, however, a different discourse has gained momentum, which is a discourse based on resilience. It corresponds to real changes in aid that were enabled by technological innovations, such as the use of digital payment systems enabling the service of populations at a distance, but there has been an especially major change in the stories that international actors tell about the nature of crises, crisis-affected populations, and their societies, and ultimately about aid itself.

The resilience paradigm rests on the notion that people, communities, and societies (can) have the capacity to adapt to or spring back from tragic life events and disasters. Disaster, rather than being a total and immobilising disruption, can become an event in which people seek continuity by using their resources to adapt. Resilience humanitarianism began in the realm of disaster relief, whereby the resilience of local people and communities and the importance of local response mechanisms became the core of the Hyogo Framework for Action in 2004. National players now take greater control of disaster response, a shift anchored in the recognition of the resilience of people and communities. International aid has retreated except in the case of mega-disasters.

Resilience humanitarianism has spilled over into other humanitarian concerns, including conflict and refugees. The refugee camp as an icon of aid is giving way to a notion that refugees are resilient in finding ways to survive.

Resilience, governance, and vulnerability

The shift towards resilience has major implications for humanitarian governance – relations between aid providers – that complicate the question of who decides about vulnerability and eligibility to aid. The classic tale of humanitarianism centred around international agencies that were assumed to be independent in their programming choices, even though the realities of aid provision involved many different types of actors. In recent years, humanitarians not only acknowledge but also espouse a shift towards much more variegated governance arrangements. They refer now to the humanitarian ecosystem, rather than the humanitarian system. National and local authorities, affected communities, and civil society are explicitly part of this ecosystem, and so are hitherto unusual humanitarian players such as the World Bank. Rather than viewing humanitarianism as a separate form of intervention, the 2016 World Humanitarian Summit proclaimed the need to bridge humanitarian action to development and peacebuilding (Ban 2016). The question of how vulnerability and eligibility are determined therefore increasingly becomes a question of whose discourse dominates, who manages the information and who makes decisions.

The shift to resilience humanitarianism also has direct implications for vulnerability and eligibility – the relation between the do-gooders and their 'beneficiaries'. A key tenet of the new way of thinking of resilience is that crisis response is much more effective and cost-efficient when it takes into account people's capacity to respond, adapt and bounce back, coined 'the resilience dividend' by the president of the Rockefeller Foundation (Rodin and Maxwell 2014). Considerable attention is given to the resilience of refugees, with literature and policy briefs converging in their portrayal of refugees as economic agents (Betts et al. 2014; Betts and Collier 2017). This leads to a form of 'resilience humanitarianism' that 'responsibilises' refugees to govern and enable their own survival (Ilcan and Rygiel 2015). Today's 'policy speak' builds on continuity between crisis and normality, and UN reports now often refer to 'crisis as the new normality'. Furthermore, resilience humanitarianism can be recognised in recent international refugee policies, in particular the Comprehensive Refugee Response Framework (United Nations 2016) and the Global Compact on Refugees (United Nations 2018). Whereas the language of resilience takes over humanitarian action, this does not mean that the corollary term of vulnerability has been abandoned. There is always a residual group of people who cannot be resilient and need to be targeted for direct assistance. Vulnerability, then, becomes a status of eligibility that is increasingly scarce. In an era of resilience humanitarianism, being a refugee is no ticket for aid eligibility anymore, and additional vulnerabilities need to be established before aid can be provided. Rather than considering categories of people (such as refugees) as vulnerable or entitled to services, individual properties must be examined before assistance is given (e.g. to *very* vulnerable refugees).

The next two sections will elaborate on these two changes: changes in governance relations that lead to complexities in defining vulnerability, and changes in aid relations that lead to restricting access to the 'vulnerability status'.

Aid governance in the disaster–conflict nexus

The increasing variation in humanitarian governance makes it difficult to identify common global discourses and practices around vulnerability. It is therefore fruitful to theorise and analyse such discourses and practices at an intermediate, meso-level for contexts that bear resemblance to each other in key aspects. The power of intermediate analysis of humanitarian praxis will be exemplified for cases where disasters meet conflict. Humanitarian action in these situations needs to improvise and navigate the conditions of operations when disaster occurs. Although it could be argued that every single case is unique, it is possible to distinguish different scenarios where certain cases are grouped together, namely high-intensity, low-intensity, and post-conflict scenarios. The following sections introduce the disaster–conflict nexus, elaborate on the idea of scenarios, and then identify for each of the scenarios how humanitarian actors deal with coordination, local actors, and vulnerability and resilience.

The disaster–conflict nexus

Every year, there are typically around 400 disasters triggered by natural hazards, mostly in lower and middle-income countries. A large number of these strike in countries affected by conflict (Peters and Budimir 2016). However, in the academic and policy fields looking at disaster, humanitarian aid, or conflict, little attention has been paid to the *nexus between disaster and conflict*. There is evidence that conflict areas are disproportionally struck by disaster. Spiegel et al. (2007) revealed that 90 per cent of conflict areas experienced one or more disaster triggered by natural hazards, a finding that corresponds with analyses of 140 events from 1998 to 2002 (Buchanan-Smith and Christoplos 2004). From 2004 to 2014, 58 per cent of disaster-related casualties were in the 30 most fragile states (Peters and Budimir 2016). There is also evidence that the impact of disasters is intensified in conflict-affected situations. Most deaths caused by disasters occur in conflict-affected and fragile states (Peters 2017), and the impact of a disaster on people's livelihoods is greater in conflict-affected and fragile contexts (Hilhorst 2013; Wisner 2012).

Conflicts are obviously caused by social processes, and this is also true for disasters. Research on disaster has overwhelmingly confirmed that the disaster outcomes of natural hazards result from processes in the socio-political context (Blaikie et al. 1994). Social processes generate unequal exposure to risk by making some people more vulnerable to disaster than others, and these inequalities are largely a function of the power relations operative in every society (Hilhorst and Bankoff 2004). Conflict can compound vulnerability and weaken the response capacities of people and communities (Wisner 2012). It also complicates responses to disasters, such as international or national relief programmes, and makes it extremely complicated to pursue disaster risk reduction activities (Mena and Hilhorst 2020).

Whereas conflict affects disasters, disasters are also seen to increase conflict risks or alter the dynamics of conflict. In a 185-country study spanning almost three decades, Brancati (2007) showed that earthquakes increased the likelihood of conflict, especially in low-income countries with a history of conflicts (see also Nel and Righarts 2008). Disasters tend to aggravate the military, socio-political and socio-economic effects of conflict (Billon and Waizenegger 2007). They can affect the military balance between parties, and relief operations can prompt additional military engagement from within or outside the country (Frerks 2008). Many conflicts have evolved or deepened because disasters evoked social protest or led to social-political change (Pelling and Dill 2010; Drury and Olson 1998).

Conversely, there is also evidence that disasters can have positive effects on conflict prevention, resolution, peacebuilding, or related processes. The differentiated effects of the Asian tsunami of 2004 have generated a great deal of scholarship on this issue (De Alwis and Hedman 2009;Gamburd 2010; Billon and Waizenegger 2007; Waizenegger and Hyndman 2010; Gaillard, Clavé, and Kelman 2008). While conflict in Sri Lanka intensified following the event, the tsunami accelerated the peace process in Aceh. A body of literature has developed around the idea of disaster diplomacy, exploring how disaster-related activities can induce cooperation between enemy parties at the national or international level (Renner and Chafe 2007; Kelman 2011).

Two major threads running through the disaster and conflict literature point to the importance of context and response. Firstly, the effects of disasters are related to previous conflict histories, especially the resilience of a state's institutions to crisis (Omelicheva 2011). Secondly, how disasters are handled mediates their effects on conflict. The impacts disasters have on conflict and stability depend on the way a government responds (Olson and Gawronski 2003; Ahrens and Rudolph 2006), to which we may add that this also depends on the responses of other actors, including the humanitarian sector. This is especially the case when we consider how conflict unfolds at different scales, showing different but inter-related dynamics. Community-level conflicts are influenced by conflict at other levels but have their own dynamics. A seven-country survey among stakeholders of the Partners for Resilience programme found that competition over natural resources and social inequality intentions are often among the factors creating disaster risks and that it is likely that these factors are further affected by disaster response, putting an additional burden on response programmes to do no harm (Hilhorst et al. 2020b).

Dealing with context: different conflict scenarios

From 2016 to 2020, the author coordinated a research team of the When Disaster Meets Conflict research programme.[2] The existing literature on disaster governance offered limited insights into the connections between disaster and conflict: studies consisted either of large-N statistical studies that treated conflict as a single category, or single case studies that treated conflict as wholly context-specific. In other words, it tended either to lump all conflicts together – whether looming conflict, full-blown war, or places where peace agreements had been signed – or it was so specific as to make it difficult to generalise the findings and be relevant to wider audiences (van Voorst and Hilhorst 2017). To deal with this conundrum, the programme worked with case-based scenarios.

The contextual realities of conflict are very diverse (Demmers 2012). The programme thus had to refine what the conflict cases represented, or as Ragin and Becker (1992) asked: what would our cases be cases of? For this, the researchers chose to embed their cases in broader 'scenarios'. In science, scenarios are usually associated with the construction of possible future developments in a known situation (Hajer and Pelzer 2018). However, scenarios are also used in training and education. For example, scenarios feature in disaster or emergency management drills in which people simulate the circumstances of an emergency so that they have an opportunity to practise their responses, ranging from fire alarm-drills in schools to multi-day exercises with detailed simulations of terrorist attacks. In these drills, participants are presented with a type of emergency and are continuously fed new pieces of information derived from a complex scenario that the organisers have assembled from real cases.

Case-based scenarios are thus theoretically constructed yet have the advantage that they maintain a realistic level of complexity and context-specificity. In doing so, they can enable communities of policy and practice to reflect on their assumptions, decisions, and impact in these types of settings. The programme constructed three conflict scenarios – high-intensity, low-intensity, and post-conflict – in a similar vein. The scenarios are researcher constructs, combining empirical properties with analytical lenses, and as such 'concerned as much with creating usable "mental models" as it is with reflecting reality' (Wood and Flinders 2014: 153). The aim is analytical and empirical generalisation concerning core disaster processes in a specific type of conflict and disaster governance.

Our three scenarios built on a small number of cases enabled an analysis of disaster governance and could connect similar cases for the sake of policy and practice. Nine country cases, three for each scenario, were selected. Each scenario involved four months of qualitative fieldwork in three case countries. The countries were South Sudan, Afghanistan, and Yemen for the high-intensity conflict scenario (Mena 2020); Ethiopia, Myanmar, and Zimbabwe for low-intensity conflict (Desportes 2020); and Nepal, Sierra Leone, and Haiti for post-conflict (Melis 2020). As the work of analysing disaster response in conflict cases progressed, the programme used the findings for diverse research uptake activities that aimed to make communities of policy and practice aware of key aspects of different conflict scenarios, associated dilemmas and decisions, and their consequences. Through these research uptake activities, feedback loops were created, and insights from practice were iteratively integrated in the scenarios. Adopting this approach helped strike a balance between case-specific contextualisation and generalisation and facilitated the uptake of findings among communities of policy and practice. The latter could use the research to fine-tune disaster response in specific types of conflict.

	Violence	*Authorities*	*Aid dynamics and major challenges for aid actors*	*Local institutions*
High-intensity conflict scenario	Violence widespread, dynamic, but uneven in space and time Casualties: 1000+ per year Primary weapons: physical violence High involvement of government in conflict	National government has reduced or no effective control over (large parts of) the country Complex governance arrangements and de-facto authorities	International actors restricted by security situation Hard to get access to populations: solutions sought in remote control aid No guarantee to monitoring and accountability	Economic crisis or disruption Increased poverty and vulnerability of local populations due to long-term state neglect and violence

Low-intensity conflict scenario	Violence less intense, more sporadic or in stalemate; violence returns over extended periods of time, cycles of repression Fewer casualties Primary weapons: policies and discourses politics of inclusion and exclusion	Government functional in large parts of country Government may be involved (civil war) or outsider of conflict Voids in governmental power and space for parallel governance Competing political structures, factions and civil society groups Government responds to conflict with 'firm grip'	International actors are restricted by authoritative government Humanitarian principles hard to maintain vis-à-vis strong state Relief operations take place on the ground State structures are primary interlocutors	Structural inequalities, marginalisation and discrimination have contributed to violence
Post-conflict scenario	Political settlement has been formally or informally reached Lingering conflict Risk of renewed conflict	National government pushes 'agenda of change' Aid to 'build back better'	Support often explicitly called upon by state High density of aid actors: risk of overwhelming with aid and undermining state agenda Non-state actors balance state's agenda and capacities with formal role of state in response	Socio-economic recovery may be uneven; humanitarian needs continue long after conflict Weak institutions, poverty and large availability of weapons may lead to high crime levels

Source: Hilhorst et al. 2019

Differentiating humanitarian discourse

Given the variety in conflict, it is not surprising that humanitarian discourses and approaches likewise vary in these different scenarios. Moreover, dealing with disaster in a conflict situation is essentially a matter of improvising, as there are no international policies available. Response models to disaster typically assume that there is a functioning government to deal with disaster response, and international policy guidelines generally do not provide guidelines on how to deal with the conflict–disaster nexus. International Humanitarian Law focuses exclusively on conflict, whereas the standing guidelines on disasters, the Hyogo Framework for Action (UN/ISDR 2007), and the Sendai Framework for Disaster Risk Reduction (United Nations 2015) do not refer to conflict. Disaster response policies are thus of little help to deal with disaster situations where the government is incapable or unwilling to act, and where the risk of aggravating conflict is high. There are marked differences between the approaches in the different scenarios. These differences could partly be explained by the diversity in conditions on the ground but also have to do with how discourses shape the ways in which humanitarians act.

Decision-making and coordination

With regard to decision-making and coordination, it might be expected that this is informed by assessments of vulnerability and resilience, but the reality is much more complex. Deciding whether to respond, where, with whom, and for whom is socially negotiated between multiple aid and society actors at different levels (national, institutional, and local), and 'real' disaster governance evolves from these processes. In high-intensity conflict (HIC) settings, the state is usually contested, and stakeholders feel legitimised to circumvent it. This does not mean that agencies follow their needs assessments. Aid action tends to be locked into path-dependent programming. Agencies tend to stay and work in the same areas and sectors over time, rather than moving to locations where aid is needed most. There are many factors that play into this, including operational challenges, inflexibility of humanitarian financing, the influence of local actors, and the roles of private companies in aid delivery (Mena 2020). Low-intensity conflict (LIC) settings often have a strong state with authoritarian tendencies where the state effectively determines what happens, often to the detriment of minorities or the regions that are home to political opposition. Governance structures are characterised by significant levels of state control and apparent collaboration between multiple aid and state actors. Tensions abound under the surface, however, with aid actors navigating bureaucracy and aiming to service people in need while avoiding confrontation with the state (Desportes 2020). In post-conflict (PC) settings, disaster response gets intertwined with the objectives and programmes of state-building. Aid actors tend to align with objectives of state-building and seek to validate the central role of the state. At the same time, they bypass state aid actors at different levels because they perceive the state to have limited capacities for coordination and implementation. Tensions often abound between disaster response delivered by humanitarians and ongoing development programming (Melis 2020).

Implementing arrangements

Decisions on whom to target for assistance are layered. They are to some extent taken at central levels, yet the nitty-gritty of who gets assistance is also largely decided by the implementing (national) agencies and/or local authorities. In the HIC scenario, fractured national governance systems result in a scattering of largely autonomous regional and local level systems of governance. Aid actors may not always operate through these systems, finding it challenging to fully understand and navigate evolving 'real' governance arrangements. This is further impeded when international political factors, including anti-terrorist legislation, prevent aid actors from working with armed opposition groups. In the HIC scenario, local actors often implement a large part of the response but are not part of central decision-making. They provide and deliver the vast majority of humanitarian and disaster aid, including disaster risk reduction (DRR), but have little or no say over funding and coordination, resulting in contradictory approaches to targeting assisted populations. In the authoritarian LIC scenario, collaboration with local actors is often centrally controlled through legal and bureaucratic regulations. LIC dynamics often mean that only civil society actors that align with the state are acceptable, while the space for others, especially those working with or advocating support for ethnic or religious minority groups, is restricted. Local actors face great difficulties working in a restricted civil space, yet this is often framed by international actors as 'local actors lacking capacity'. In the PC scenario, tensions abound between different levels and domains of the state, each seeking to expand its mandate and financial power. The central state is often far removed from the affected populations and local authorities. Aid actors may find themselves subject to the push and pull of intra-state competition.

Hence, in all three scenarios, in different ways, major cleavages within aid-state and central-local aid relations impede bottom-up informed choices and decisions regarding vulnerability, resilience, and targeting. Trust is a major issue. Humanitarian action in HIC scenarios is often cloaked in the language of 'remote management'. This means that local actors face the most serious security risks (the ethics of 'outsourcing' security risks is a major issue), while central actors remain insecure about their motivations and allegiances and hence seek to remotely control all their actions. In the LIC scenario, local actors are often represented as biased and partisan, even when raising legitimate concerns about the rights and needs of communities. In the PC scenario, national and local NGOs are considered implementing partners but are not always accepted in state-aid coordination mechanisms. Frames abound among centrally positioned humanitarians about the weak capacities or corruption of both local NGOs and state institutions, while ignoring comparable problems with their own integrity and relying on a limited definition of 'capacity'.

Politicisation and instrumentalisation

Disaster response inevitably becomes part of the politics of conflict. Actors use the disaster in their struggle for control and legitimacy. The state may instrumentalise or even 'weaponise' the disaster response to achieve political goals.

It can prevent aid from reaching certain areas to weaken an area held by armed opposition groups (as seen in HIC settings), or further marginalise a minority group (as seen in LIC settings). In the PC scenario, disaster response and state-building intertwine. Disaster response can play into – helpfully or not – legislative processes (for instance, accelerating the new constitution in Nepal) or can be exploited for electoral gain, as was seen in Haiti and Sierra Leone. On the other hand, disaster response can also de-escalate conflict dynamics, such as through DRR programmes in Afghanistan, or be framed as a neutral and technocratic space enabling collaboration, such as in the LIC contexts of Ethiopia and Myanmar. In the LIC scenario, non-state actors find it difficult to openly challenge state-led response systems. Most non-state actors opt for a non-confrontational, self-censoring approach – navigating around challenges rather than tackling them head-on and refraining from speaking out. In doing so, aid actors run the risk of reinforcing power imbalances as well as contributing to shrinking humanitarian and civil society space and effectively ignoring the needs of communities unfavoured by the state (Desportes and Hilhorst 2020).

Resilience and vulnerability

The language of resilience and vulnerability is likewise politicised and instrumentalised. In HIC scenarios a language of resilience is often used, when in fact it is the inability of aid agencies to reach people in need that matters. Labelling areas out of reach as resilient is comforting for aid actors (Walkup 1997, Jaspars, this volume) and convenient for policymakers. In LIC scenarios, definitions of resilience and vulnerability are to some extent subject to the politics of national authorities. Authoritarian governments tend to downplay vulnerabilities in order to maintain their appearance of control. Aid agencies need to self-censor their reports, for example, avoiding the word 'cholera' where this is deemed unacceptable, or becoming dependent on the information made available by the government on harvests and nutrition (Desportes et al. 2019). In PC scenarios, it was often found that central aid actors frame populations as resilient while maintaining a strong discourse about institutional vulnerability vis-à-vis state and non-state actors.

The importance of framing: refugees in and around Europe

Two conclusions can be derived from the scenarios of disaster and conflict. First, general trends in resilience and vulnerability are translated very differently in scenarios with differing politics and governance. Determining the eligibility for humanitarian assistance is not just in the hands of humanitarians, but highly depends on other actors and factors. Second, the way in which resilience and vulnerability is determined is only partly related to actual conditions of need. They are also derived from images or frames that decisive actors have about situations and people. This can be clearly illustrated with yet another scenario, concerning the situation of migrants that came in large numbers to Europe in 2015 (Hilhorst et al. 2020a, Jaspars and Hilhorst 2021). The discursive reading of the situation in this case is highly contradictory, starting with the terms used for the crisis and people involved. Politicians tend to talk about the 'European migration crisis', disregarding the fact that most of the migrants came from war-affected countries and should be labelled as refugees. Humanitarians therefore tended to talk about a 'refugee crisis'. Volunteers and activists, on the other hand, pointed out that the crisis might be more aptly labelled as a 'solidarity crisis'. These frames are void of references to the actual vulnerability of the people involved. Many of the people trying to come to Europe have mixed motivations, from running away from war to finding better life chances elsewhere. By the time they arrive at European borders, however, their destitution and trauma (whether rooted in their country of origin or their journey – e.g. encountering violence, sexual exploitation and death in crossing the Mediterranean) has usually brought about a high state of vulnerability, which seems to be of no consequence for their entitlement to assistance or refuge. The politicisation of resilience and vulnerability, in this scenario, centres on ideas of deserving and undeserving migrants, where migrants as well as the people daring to come to their assistance are subject to criminalisation (Pusterla 2021).

Everyday politics of vulnerability

Whereas shifts in humanitarian governance and discourse have profound implications for eligibility of aid, it remains important to scrutinise how humanitarian agencies deal with vulnerability in practice. A lot of attention has focused on the shift towards resilience; however, the flipside of resilience continues to be vulnerability. It is vulnerability that dictates where assistance gets allocated. Vulnerability is increasingly invisible in resilience humanitarianism and there is little attention to the technologies by which vulnerability is determined. Major questions about this can be derived from Science and Technology Studies. They concern historical and contemporary production, embedded forms of knowledge, and the organisation of humanitarian technologies and their impact on vulnerable people. The last section of the chapter will discuss three issues of everyday politics that are pertinent to the identification of vulnerability and will need further research in the years to come (Sandvik et al. 2014).

Manipulating indicators of vulnerability

People's eligibility for assistance depends on their vulnerability, but what happens when too many people are vulnerable? This conundrum is all too familiar to humanitarians, who often find themselves in situations where vulnerability outweighs their ability to respond. Susanne Jaspars (this volume) provides evidence on how humanitarian agencies in Sudan maintain a story of resilience, even though it is unfounded, to fabricate a story line of diminished vulnerability that can legitimise the reduction of aid rations. A striking example of manipulating indicators of vulnerability was related by an aid worker in Lesbos, who told the author that:

> After the Turkey deal, UNHCR had to establish the vulnerability of people to decide who could be transferred to the mainland of Greece and in turn assign a procedure. When they found out that almost all residents of Moria would fit the criteria, a new category was invented of the *very vulnerable*. Only these very vulnerable, with much stricter criteria, became eligible for transfer to the mainland.[3]

The author was not able to confirm this with UNHCR, but the concept of *very vulnerable* is often used in humanitarian reports. Humanitarian advocacy, as found in countless reports, calls upon donors to provide more funding in view of the widespread vulnerability of their populations of concern. A recent report by the Danish Refugee Council, for example, provides ample evidence on how COVID-19 has increased the vulnerability of Syrian refugees in Lebanon, Jordan, and Turkey. While advocating support for vulnerable people, they make nonetheless a distinction between vulnerable and very vulnerable people. It is not clear what the difference is between these two categories, and hence it remains unclear how agencies manage eligibility and decide on whom to assist, and what this means for people who are not being assisted because they are merely vulnerable instead of *very* vulnerable (DRC 2021).

Categorisation

A second issue concerns the everyday politics of categorisation. Despite the idea of humanitarian assistance to focus on vulnerabilities at an individual level, questions of scale and timeliness drive humanitarian action to work with preconceived categories in practice. However, these categories may incorporate implicit bias. One example is the conception of disabled people as vulnerable. A refugee I met in Turkey in 2016, was moving around with difficulty in a wheelchair, paralysed from the waist down. He had found it relatively easy to get access to a monthly relief payment. It was not enough to sustain his family but at least enabled him to rent a room where the family could stay. On the other hand, people with less visible problems, such as disease or mental health issues, faced much more difficulty in obtaining such access.

Ascribing categorical vulnerability is nowhere more apparent than in the domain of gender (see also Bradshaw, this volume). Charli Carpenter (2005) coined the one-word notion of 'womenandchildren', to express how women are automatically boxed into vulnerability categories together with children. In the Syria crisis, many relief programmes were only eligible for women. Men were expected to be able-bodied and self-reliant. However, without a work permit and without protection against exploitation, men could find themselves destitute.

This categorisation of vulnerability was related to responsibilities of care that many women have and was reinforced by imagery depicting women as innocent and deserving of our compassion. Men, on the other hand, represented danger. They were seen as a danger to their partners, whom they beat, oppress or rape, and as a danger to society, as they were associated with violence or even terrorism. However, the vast majority of (young) men who run away from violence are vulnerable victims. Men can be forcibly recruited, tortured, molested, shot, or castrated, but their involvement in psycho-social support projects is rare. When these projects reach out to men, it is usually to engage them for the cause of women and make them advocates to stop violence against women. While the number of women abused by men is staggering in many conflict-affected societies, this should not lead to lumping all men together and excluding them from the humanitarian gaze.

Using algorithms to identify the vulnerable

Another area of technology begging more in-depth research is the use of algorithms in humanitarian action. Deciding whom to assist with cash relief increasingly relies on algorithms and, as Paul Currion (2016) cautioned, 'these algorithms are not just powerful – they're also opaque'. The use of cash relief has soared in humanitarian action; it is seen as more dignified and vastly more effective than traditional forms of food distribution, and it does not centre on camps where people are concentrated to enable their survival. Whereas a lot has been written about its advantages (Harvey 2007, Heaslip et al. 2016), the nitty-gritty of how families are selected for relief can be opaque, and stories abound of people that are perplexed that their neighbours received cash relief and they didn't, even though they live in similar circumstances.

The same issues apply to the new approach of anticipatory humanitarian action. This means that funds usually reserved for response can be released before a disaster happens when an impact-based forecast – i.e., the expected humanitarian impact as a result of the forecasted weather – reaches a predefined danger level. Anticipatory action is potentially pathbreaking and is well-suited to resilience humanitarianism. A review states: 'There is a growing consensus that anticipatory humanitarian action, where it is possible, can be faster and cheaper. And because it empowers people to protect themselves on their own terms, it is more dignified' (Moser 2021). The danger of technology-driven assistance and possible accountability deficits beg scrutiny in the years to come (Bierens et al. 2020). As Currion (2016) noted, it is the responsibility of humanitarians 'to make sure that the algorithms reflect their values, rather than leaving key moments of micro-decisions on the criteria of inclusion to app developers'.

Another concern about the automation of selecting the most vulnerable for assistance is about privacy and the use of personal data. In order to avoid cheating with cash relief, humanitarian agencies increasingly use biodata, for example, through an eye scan. This requires the highest standards of data protection, but a recent scandal showed this may not be the case. Investigations of Human Rights Watch revealed that UNHCR has shared biodata of 850,000 Rohingya refugees with the Government of Myanmar, the regime responsible for their expulsion from the country. The consequences of this are yet to unfold (HRW 2021).

Conclusion

This chapter has reviewed issues pertaining to the use of the concept of vulnerability in humanitarian action. We live in an era of resilience humanitarianism, and at first sight it seems that vulnerability has been relegated to the back burner. In an increasingly scattered and changing humanitarian landscape, with different constellations of governance, multiple sets of ethics, and extensive accountability relations, it is important to analyse how humanitarian actors deal with vulnerability.

Firstly, it has been elaborated how shifts in humanitarian governance complicate the question of who decides about vulnerability and eligibility to assistance. Using case-based scenarios (such as high-intensity conflict, low-intensity conflict, post-conflict, or European migration) may underpin in-depth insight into how humanitarian actors deal with resilience and vulnerability under specific conditions. Secondly, it has called attention to the importance of the everyday politics of humanitarian technology. In the nitty-gritty of how vulnerability conceptions are applied in practice, it is found how humanitarian action can affect people's chances to survive and co-shape the relations between them. Humanitarians become increasingly strict in recognising vulnerability with reference to resilience. In doing so, they risk abandoning large numbers of people whose vulnerability is the only entitlement they have for minimal survival.

Notes

1 This chapter is based on research of two programmes, and I gratefully acknowledge their contributions. These are NWO, Netherlands Organisation for Scientific Research, VICI no. 453/14/013 and the European Research Council (ERC) under the European Union's Horizon 2020 research and innovation programme, grant agreement No 884139.

2 https://www.iss.nl/en/research/research-projects/when-disaster-meets-conflict

3 Interview aid worker, July 2019.

References

Ahrens, Joachim, and Patrick Rudolph. 2006. "The importance of governance in risk reduction and disaster management". *Journal of Contingencies and Crisis Management*, 14(4), 207–220. doi:10.1111/j.14685973.2006.00497.x

Ban, Ki-Moon. 2016. "One humanity: shared responsibility Report of the Secretary-General for the World Humanitarian Summit." UN General Assembly (accessed February 2 2022).

Betts, Alexander, Louise Bloom, Josiah Kaplan, and Naohiko Omata. 2014. *Refugee Economies: Rethinking Popular Assumptions*. Oxford: Humanitarian Innovation Project,

University of Oxford. Available at www.rsc.ox.ac.uk/files/files-1/refugee-economies-2014.pdf.

Betts, Alexander, and Paul Collier. 2017. *Refuge: Transforming a Broken Refugee System*. London: Penguin.

Bierens, Sterre, Kees Boersma, and Marc JC van den Homberg. 2020. "The legitimacy, accountability, and ownership of an impact-based forecasting model in disaster governance." *Politics and Governance*, 8(4), 445–455.

Blaikie, Piers, Terry Cannon, Ian Davis, and Ben Wisner. 1994. "At risk", *Natural Hazards, People's Vulnerability and Disasters*. London: Routlegde.

Brancati, Dawn. 2007. "Political aftershocks: The Impact of earthquakes on intrastate conflict". *Journal of Conflict Resolution*, 51(5), 715–743. doi:10.1177/0022002707305234.

Buchanan-Smith, Margie, and Ian Christoplos. 2004. "Natural disasters amid complex political emergencies". *Humanitarian Exchange*, 27, 36–38.

Carpenter, R. Charli. 2005. ""Women, children and other vulnerable groups": Gender, strategic frames and the protection of civilians as a transnational issue". *International Studies Quarterly*, 49(2), 295–334.

Currion, Paul. 2016. "Slave to the algorithm". *New Humanitarian*. https://www.thenewhumanitarian.org/opinion/2016/07/11/slave-algorithm.

Danish Refugee Council. 2021. "Global COVID-19 response". https://drc.ngo/about-us/for-the-media/press-releases/2021/3/new-drc-report-on-COVID-19-response/

De Alwis, Malathi, and Eva-Lotta Hedman. 2009. *Tsunami in a Time of War: Aid Activism and Reconstruction in Sri Lanka and Aceh*. Vol. 4. Sri Lanka: International Centre for Ethnic Studies.

Demmers, Jolle. 2012. *Theories of Violent Conflict: An Introduction*. London: Routledge.

Desportes, I. 2020. *Repression Without Resistance: Disaster Responses in Authoritarian Low-Intensity Conflict Settings*. Erasmus University Rotterdam. Available at: https://repub.eur.nl/pub/131548 (accessed May 29 2021).

Desportes, I., and Hilhorst, D. 2020. "Disaster governance in conflict-affected authoritarian contexts: The cases of Ethiopia, Myanmar, and Zimbabwe". *Politics and Governance*, 8(4), 343–354.

Desportes, Isabelle, Hone Mandefro, and Dorothea Hilhorst. 2019. "The humanitarian theatre: Drought response during Ethiopia's low-intensity conflict of 2016". *The Journal of Modern African Studies*, 57(1), 31–59. doi:10.1017/S0022278X18000654

Drury, Cooper, and Richard Stuart Olson. 1998. "Disasters and political unrest: An empirical investigation". *Journal of Contingencies and Crisis Management*, 6(3), 153–161. doi:10.1111/1468-5973.00084.

Frerks, Georg. 2008. "Tsunami response in Sri Lanka: Civil-military cooperation in a conflictuous context." *Managing Civil-Military Cooperation: A*, 24(7), 67–80.

Gaillard, Jean-Christophe, Elsa Clavé, and Ilan Kelman. 2008. "Wave of peace? Tsunami disaster diplomacy in Aceh, Indonesia." *Geoforum*, 39(1), 511–526.

Gamburd, Michele Ruth. 2010. "The golden wave: Discourses on the equitable distribution of tsunami aid on Sri Lanka's southwest coast". In *Tsunami Recovery in Sri Lanka: Ethnic and Regional Dimensions* edited by Dennis McGilvray and Michele Ruth Gamburd, 64–84. London: Routledge.

Hajer, Maarten, and Peter Pelzer. 2018. "2050—An energetic odyssey: Understanding 'Techniques of Futuring' in the transition towards renewable energy". *Energy Research and Social Science*, 44, 222–231. doi:10.1016/j.erss.2018.01.013.

Harvey, Paul. 2007. "Cash-based responses in emergencies". *IDS Bulletin*, 38, 3.

Heaslip, Graham, Ira Haavisto, and Gyöngyi Kovács. 2016. In *Advances in Managing Humanitarian Operations* edited by Christopher W. Zobel, Nezih Altay, and Mark P. Haselkorn, 59–78. Cham: Springer,

Hilhorst, Dorothea. 2013. "Disaster, conflict and society in crises: Everyday politics of crisis response". In *Disaster, Conflict and Society in Crises: Everyday Politics of Crisis Response* edited by Dorothea Hilhorst, 1–15, New York: Routledge, humanitarian studies series 1.

Hilhorst, Dorothea. 2018. "Classical humanitarianism and resilience humanitarianism: Making sense of two brands of humanitarian action." *Journal of International Humanitarian Action*, 3(1), 1–12.

Hilhorst, Dorothea, and Greg Bankoff. 2004. In *Mapping Vulnerability: Disaster, Development and People* edited by Greg Bankoff, Georg Frerks and Dorothea Hilhorst, 20–28. London: Earthscan.

Hilhorst, Dorothea, Maria Hagan, and Olivia Quinn. 2020a. "Reconsidering humanitarian advocacy through pressure points of the European 'migration crisis'". *International Migration*, 59(3) (2021): 125–144.

Hilhorst, Dorothea, Rodrigo Mena, Roanne van Voorst, Isabelle Desportes, and Samantha Melis. 2019. Disaster risk governance and humanitarian aid in different conflict scenarios. Contributing Paper to GAR 2019 65903_f303hillhorstdisasterresponseandhum.pdf (preventionweb.net).

Hilhorst, Dorothea, Marie José Vervest, Isabelle Desportes, Samantha Melis, Rodrigo Mena, and Roanne van Voorst. 2020b. *Strengthening Community Resilience in Conflict: Learnings from the Partners for Resilience Programme*. Partners for Resilience. https://library.partnersforresilience.nl/?r=525.

Human Rights Watch. 2021. "UN Shared Rohingya Data Without Informed Consent". Published 15 June at https://www.hrw.org/news/2021/06/15/un-shared-rohingya-data-without-informed-consent.

Ilcan, Suzan, and Kim Rygiel. 2015. "'Resiliency Humanitarianism': Responsibilizing refugees through humanitarian emergency governance in the camp". *International Political Sociology*, 9(4), 333–351.

Jaspars, Susanne, and Dorothea Hilhorst. 2021. "Introduction 'Politics, humanitarianism and migration to Europe'". *International Migration*, 59(3), 3–8.

Kelman, Ilan. 2011. *Disaster Diplomacy: How Disasters Affect Peace and Conflict*. London: Routledge.

Billon, Philippe Le, and Arno Waizenegger. 2007. "Peace in the Wake of Disaster? Secessionist Conflicts and the 2004 Indian Ocean Tsunami". *Transactions of the Institute of British Geographers*, 32(3), 411–427.

Melis, Samantha. 2020. *Constructing Disaster Response Governance in Post-Conflict Settings: Contention, Collaboration and Compromise*. Erasmus University Rotterdam. Available at: https://repub.eur.nl/pub/134590 (accessed May 29 2021).

Mena, Rodrigo, and Dorothea Hilhorst. 2020. "The (im)possibilities of disaster risk reduction in the context of high-intensity conflict: The case of Afghanistan". *Environmental Hazards*, doi:10.1080/17477891.2020.1771250.

Mena, Rodrigo. 2020. *Disasters in Conflict: Understanding Disaster Governance, Response, and Risk Reduction During High-Intensity Conflict in South Sudan, Afghanistan, and Yemen*. Erasmus University Rotterdam. Available at: https://repub.eur.nl/pub/134732 (accessed May 24 2021).

Moser, Patrick. 2021. Acting Before the Flood. An Anticipatory Humanitarian Action Pilot in Bangladesh. OCHA, March OCHA (2021) Acting Before the Flood-Bangladesh AA synthesis report.pdf.

Nel, Philip, and Marjolein Righarts. 2008. "Natural disasters and the risk of violent civil conflict". *International Studies Quarterly*, 52(1), 159–185. doi:10.1111/j.1468-2478.2007.00495.x.

Olson, Richard, and Vincent Gawronski. 2003. "Disasters as critical junctures? Managua, Nicaragua 1972 and Mexico City 1985". *International Journal of Mass Emergencies and Disasters*, 21(1), 5–35.

Omelicheva, Mariya. 2011. "Natural disasters: Triggers of political instability?". *International Interactions*, 37(4), 441–465. doi:10.1080/03050629.2011.622653.

Pelling, Mark, and Kathleen Dill. 2010. "Disaster politics: Tipping points for change in the adaptation of sociopolitical regimes". *Progress in Human Geography*, 34(1), 21–37. doi:10.1177/0309132509105004.

Peters, Katie. 2017. *The Next Frontier for Disaster Risk Reduction*. October. London, UK: ODI (Overseas Development Institute). Retrieved 15 October 2017 from https://www.odi.org/publications/10952-next-frontier-disaster-risk-reduction-tackling-disasters-fragile-and-conflict-affectedcontexts.

Peters, Katie, and Mirianna Budimir. 2016. "When disasters and conflict collide: Facts and figures". ODI (Overseas Development Institute). https://www.odi.org/publications/10410-when-disasters-andconflicts-collide-facts-and-figures.

Pusterla, Francesca. 2021. "Legal perspectives on solidarity crime in Italy." *International Migration*, 59(3), 79–95.

Ragin, Charles, and Howard Becker. 1992. *What is a Case? Exploring the Foundations of Social Inquiry*. Cambridge, UK: Cambridge University Press.

Renner, Michael, and Zoë Chafe. 2007. *Beyond Disasters: Creating Opportunities for Peace*. Worldwatch Report, Washington DC: Worldwatch Institute.

Rodin, Judith, and Cyndee Maxwell. 2014. *The Resilience Dividend*. London, UK: Profile Books.

Sandvik, Kristin Bergtora, Maria Gabrielsen Jumbert, John Karlsrud, and Mareile Kaufmann. 2014. "Humanitarian technology: A critical research agenda." *International Review of the Red Cross*, 96(893), 219–242.

Spiegel, Paul B., Phuoc Le, Mija-Tesse Ververs, and Peter Salama. 2007. "Occurrence and overlap of natural disasters, complex emergencies and epidemics during the past decade (1995–2004)." *Conflict and Health*, 1(January), 2. doi:10.1186/1752-1505-1-2.

UN/ISDR. 2007. *Hyogo Framework for Action 2005-2015: Building the Resilience of Nations and Communities to Disasters*. Geneva: United Nations Office for Disaster Risk Reduction (UNISDR). https://www.unisdr.org/files/1037_hyogoframeworkforactionenglish.pdf.

United Nations. 2015. *Sendai Framework for Disaster Risk Reduction*. Sendai, Japan: UNISDR. http://www.unisdr.org/we/coordinate/sendai-framework

United Nations. 2016. Comprehensive Refugee Response Framework https://www.unhcr.org/comprehensive-refugee-response-framework-crrf.html (accessed December 2020].

United Nations. 2018. "Global Compact On Refugees". https://www.unhcr.org/5c658aed4.pdf (accessed December 2020).

Van Voorst, Roanne, and Dorothea Hilhorst. 2017. *Humanitarian Action in Disaster and Conflict Settings. Insights from an Expert Panel*. The Hague, Institute of Social Studies. Available at: www.iss.nl/news_events/iss_news/detail_news/news/11526-human-action-in-disaster-and-conflict-settings (accessed May 20 2021).

Waizenegger, Arno, and Jennifer Hyndman. 2010. "Two solitudes: Post-tsunami and post-conflict Aceh." *Disasters*, 34(3), 787–808.

Walkup, Mark. 1997. "Policy dysfunction in humanitarian organizations: The role of coping strategies, institutions, and organizational culture." *Journal of Refugee Studies*, 10(1), 37–60.

Wisner, Ben. 2012. "Violent conflict, natural hazards and disaster". In *The Routledge Handbook of Hazards and Disaster Risk Reduction* edited by Ben Wisner, J.C. Gaillard, and Ilan Kelman, 65–76. London: Routledge.

Wood, M., and M Flinders. 2014. "Rethinking depoliticisation: Beyond the governmental". *Policy and Politics*, 42(4), 151–170.

8

RESILIENCE, FOOD SECURITY, AND THE ABANDONMENT OF CRISIS-AFFECTED POPULATIONS[1]

Susanne Jaspars

Introduction

In 2004, the United Nations Humanitarian Coordinator for Sudan called Darfur the world's worst humanitarian crisis (BBC 2004). This was soon followed by the World Food Programme's (WFP) largest food aid operation. Yet fifteen years later, whilst conflict and violence was ongoing, international agencies were withdrawing food assistance and were unable to access many conflict-affected populations. Levels of acute malnutrition in much of Darfur were well above internationally recognised emergency thresholds but were considered a consequence of people's own actions and behaviours rather than vulnerability to ongoing attacks and other threats to livelihoods. Aid organisations encouraged resilience by promoting behaviour change and by providing services to improve health and nutrition. Under Sudan's previous government, this 'regime of practices' (Schaffer 1984) suited not only international aid agencies who had to programme remotely, but also the government because it hid their actions as the cause of ongoing crisis and vulnerability, and absolved the international community from the responsibility to protect. This was particularly convenient in a context where the EU collaborated with the Sudan government to stem migration to Europe. The situation in Sudan changed dramatically with the popular uprising and overthrow of President Bashir regime in April 2019. Despite an initial period of optimism for transition to democracy, however, in 2021 (the time of writing) the country continued to face political instability, a severe economic and humanitarian crisis, a comprehensive peace agreement has proved difficult to achieve and Darfur has seen a resurgence of violence and displacement. Aid practices remain largely the same (Jaspars and El-Tayeb 2021).

The situation in the past two decades sharply contrasts with the time I worked as a nutritionist for Oxfam in Darfur, in 1989 and 1990. At that time, aid agencies

DOI: 10.4324/9781003219453-10

could travel freely around Darfur. I travelled with my Sudanese colleagues to remote villages, where we used to stay for several days to make sure we talked to different people, observed how they lived, and could interpret nutritional data within that particular context. We would make recommendations for income and agricultural interventions as well as food aid to support livelihoods while recognising that the main social, political and economic causes of malnutrition were beyond the capacity of the community to address. Advocacy to bring about policy change, and representing crisis-affected populations to the Sudan government and international actors, was considered a key part of our work (Young and Jaspars 1995). This was also the time of Operation Lifeline Sudan in response to conflict and famine in southern Sudan, when the UN (with the support of Western donors) negotiated access to war-affected populations (Karim et al. 1996).

This chapter examines how such a contrast in aid practices came about in a relatively short period of time and argues that resilience practices led to the creation of a regime of truth that perpetuates and helps maintain crisis in Darfur. The chapter analyses resilience practices, in particular food-based resilience practices, as a way of governing beyond the state – governmentality in Foucault's terms (Foucault 2007). In other words, how a range of techniques, tactics, and organisations (and their underlying assumptions) influence behaviour, power relations and become a way of managing populations. Rather than only looking at what policies and institutions intended to achieve, this chapter also examines the effect of the actual regimes of practices used. This includes the production of a 'regime of truth' in which a particular discourse is accepted as truth, produced through specific tools or techniques accepted as valid at a particular point in time (Foucault 1980). The analysis also uses the work of Bernard Schaffer (1984) on public policy and David Keen (1994) on the *Benefits of Famine* –by examining what policy practices actually do. The failure of relief in response to the 1988 famine in Bahr Al-Ghazal, Sudan, for example, was a success for government counter-insurgency, and for the merchants and soldiers who made a profit out of distress sales of livestock and high grain prices (ibid.).

The chapter also builds on the resilience literature, in particular that which analyses resilience approaches as a form of neoliberal governmentality. Resilience approaches promote the creation of autonomous and responsible subjects who can adapt and survive in situations of repeated crisis or uncertainty (Welsh 2014). They shift responsibility from state or international community, to individual, which according to Joseph (2016) has been justified on the basis of failures of international interventions; i.e. it is better to focus on local capacities. This 'responsibilisation' is not only implicit in policy documents, but in a whole assemblage of discourses, institutions, architectural forms, laws, etc. (Howell 2015, 68). Furthermore, the concept of resilience can be seen as both dehumanising and depoliticising: rather than addressing vulnerability through political resistance to the conditions which produce suffering, populations are expected to 'accept the necessity of living a life of permanent exposure to endemic dangers' (Evans and Reid 2013, 13). Resilience approaches therefore expect people to reduce their own vulnerability but not to address its causes. A limited number of researchers have applied this kind of analysis to situations of humanitarian crisis, where promoting resilience has become a key aim (Levine and Mosel 2014; Scott-Smith 2018). One the one hand, it appears to offer a solution to protracted crisis through creating opportunities for adaptation, but on the other hand, these opportunities are only at the micro or everyday level (Joseph 2016; Hilhorst 2018). In refugee camps, for example, encouraging 'self-government' and entrepreneurship to survive in the face of declining levels of aid, can be seen as promoting the acceptance of the dismal conditions of the camp (Ilcan and Rygiel 2015). This element of survival within long-term displacement is also evident in the recently agreed Global Compact on Refugees and the Global Compact on Migration (UNHCR 2018; Global Compact for Migration 2018). Duffield (2016) links resilience approaches with the advance of digital humanitarianism, which enables self-organisation in the absence of social welfare and critical infrastructure. This chapter takes the analysis of resilience approaches in situations of humanitarian crisis further and demonstrates how in Darfur, they have led to the abandonment of crisis-affected populations.

While the focus of the chapter is Sudan, in particular Darfur, the argument it puts forward is relevant globally. Sudan has functioned as a laboratory for aid practices for at least 50 years, so what happens there is a good test case for the effects of aid practices generally. The chapter draws mainly on my Ph.D. research completed in 2016 (see Jaspars 2018a) but is also informed by discussions with Darfuris and literature published on resilience and on Darfur more recently. Ph.D fieldwork in 2012 and 2013 included interviews with long-term aid workers, aid agencies working in Darfur, government officials, beneficiaries, traders, and transporters. The author visited Sudan again in 2014, 2016, and 2017 for various studies and dissemination events. The chapter starts with an analysis of how resilience ideology came to dominate aid practices in Darfur. The following section analyses how food-based resilience practices have produced a regime of truth which has made conflict and power relations invisible and which enabled the withdrawal of humanitarian aid while conflict and violence are ongoing. The chapter then argues that resilience is also a fantasy, because it entails an element of denial. The final section, before conclusions, reviews Darfur's ongoing humanitarian crisis in relation to resilience ideology.

How did food-based resilience practices come about?

An overview of current food-based resilience approaches in Darfur

In 2021, detailed information on the humanitarian situation in Darfur was hard to come by, apart from overall numbers in need, yet promoting resilience is an important part of aid practices. The UN OCHA humanitarian needs overview for that year reports 13.4 million people in need of humanitarian assistance in Sudan, including 2.5 million Internally Displaced Persons (IDPs), mostly in

Darfur (UN OCHA 2021a). These days, reporting populations in need does not mean recommending assistance. Out of 13.4 million in need, 8.9 people were targeted for assistance, including 1.5 out of 2.5 million IDPs. Those in need of food security support were estimated at 8.2 million, with 6.2 million targeted; out of which 4.8 million will receive life-saving food assistance and 2.2 million other forms of resilience or protection support (UN OCHA 2021a, 2021b). Changes in discourse have gradually reduced those who receive assistance: from IDPs, to IDPs in need, to only the most vulnerable receiving assistance (see also UN OCHA 2018, 2019). At the same time, food insecurity has increased over the past three years, with deepening economic crisis, and – even after the revolution – increased violence and displacement in Darfur and the Covid-19 pandemic (Fewsnet 2021; WFP Sudan 2020). While much detail on the nature of protracted conflict and associated risks is lacking from this official information, it indicates ongoing vulnerability and crisis for large sections on the population.

In the past 10–15 years, resilience has been a key theme of humanitarian operations, including in Darfur, which has been accompanied by overall decline in material assistance (see below). Such interventions have largely focussed on behaviour change, building capacity, and treatment at the individual level. The 2021 humanitarian response plan aims for part of the food insecure and vulnerable population to receive livelihood support (for example food-or cash- for work) and basic services to enhance resilience. The plan also aims to strengthen health systems to promote people's ability to 'absorb and recover from shocks' (UN OCHA 2021b, 47). This is not the first time resilience has come up in humanitarian response planning, in 2014, for example, repair of clinics and schools, support for self-help groups, animal health and agricultural extension, hygiene promotion, psychosocial support and skills training were suggested as promoting resilience (UN OCHA 2014).

Food security and nutrition has become key in promoting resilience. In WFP's 2013 strategic plan, for example, food aid objectives include building resilience by supporting nutrition, the establishment of safety nets, and working with the private sector (WFP 2013a). This was re-affirmed in their policy on building resilience for food security and nutrition, which somewhat confusingly turns this round and highlights the importance of improving nutrition to build resilience (WFP 2015). These approaches remain important in WFP's latest strategic plan (WFP 2017) and for other food-centred UN organisations. Nutrition is understood as key to resilience because well-nourished people can work harder and are better able to withstand shocks and stresses (FAO 2012; UNICEF 2013). Food security interventions such as agricultural production support, income generation, public works, and famine early warning initiatives have all been re-invented as resilience interventions (Levine and Mosel 2014). USAID in Sudan saw cash transfers, or food vouchers, as promoting resilience by strengthening markets and dietary diversity.[2] Resilience, or rather the lack of resilience, has also been linked to migration. As with resilience approaches generally, those aimed at reducing migration are focussed on food security, nutrition and health interventions (for example the EU's interventions to tackle irregular migration and displacement for the Horn of Africa) (European Commission 2016). In Darfur, such interventions include: strengthening local health and nutrition services, and supporting water catchment systems to support livelihoods. This has not, in fact, stopped forced migration (Jaspars and Buchanan Smith 2018).

How did a resilience regime emerge?

Resilience approaches came about by changes in both global and local politics, and the perceived failures of previous livelihoods-centred approaches. Prior to the 2000s, food aid practices can be considered part of a livelihoods regime; a regime in which food aid practices aimed to save livelihoods as well as lives. Starting with the refugee crises and famines in the early 1980s, International Non-Governmental Organisations (INGOs) attempted to target emergency food aid directly at communities or individuals worst affected by famine. Later that decade, food aid was provided to support 'coping strategies' and assist populations to remain on their land and as such promote self-reliance (Buckley 1988). Livelihood support remained prominent in humanitarian interventions in the 1990s, including market, agricultural, and income support as well as food distribution and a range of new assessment and monitoring systems (Young et al. 2004). By the end of the 1990s, however, many populations – particularly in Sub-Saharan Africa – found themselves in situations of protracted crisis, suffering repeated or persistently high levels of acute malnutrition (Ockwell 1999). Newly-developed famine early warning systems had not led to early response (Buchanan-Smith and Davies 1995), and methods for targeting the most vulnerable had usually failed (Jaspars and Shoham 1999). Every single evaluation of food aid operations in Darfur in the 1980s and 1990s concluded that it had not met the objectives of supporting livelihoods, usually defined as supporting coping strategies and preventing the depletion of assets and distress migration (see, for example, Buchanan Smith 1989; Osman 1993; DfID 1997).

These experiences of the 1990s came together with heightened fears of global instability with the War on Terror (WoT), the 2008 food and finance crisis, and the prospect of further crises due to climate change. Resilience, or the ability to resist and adapt to shocks, became the way of thinking about intervening in situations of protracted crisis and ongoing risks, resulting in new initiatives to promote food security and global stability. These initiatives included a UN Comprehensive Framework for Action following the 2008 food crisis (FAO 2008) and a number of Public-Private Partnerships, which had the aim of improving access and functioning of markets and promoting greater private sector engagement, as ways of meeting immediate needs and building resilience (ibid.). One PPP initiative is the Scaling Up Nutrition (SUN) movement, established in 2010, consisting of UN, donors, INGOs, business, and scientists (SUN 2014). Its aim is to scale up a standard package of medicalised and behavioural nutrition interventions, which were recommended in a highly influential set of papers in the

Lancet in 2008 (Arnold and Beckmann 2011). They concluded that substantial reductions in malnutrition could be achieved by a basic set of interventions such as food fortification, vitamin supplementation, and nutrition education on breastfeeding and weaning practices (Black et al. 2008). A simultaneous trend in the 2000s has been the growth of Ready to Use Therapeutic Foods (RUTF) as part of Community Managed Acute Malnutrition (CMAM). The *Lancet* papers and the use of RUTF form the basis of most agency and donor guidelines on nutrition, thus leading to the widespread medicalisation of malnutrition: a focus on measurement, treatment, and education. As such, it provides a prime example of the hyperneoliberalism of the resilience regime. Despite global financial systems being a major cause of the 2008 food crisis, the response has been to 'responsibilise' individuals to change their behaviour to become more resilient, and to focus on treatment with specialised food products. RUTF and other 'nutraceuticals', sell the possibility of survival within a context of permanent emergency (Street 2015). As Schaffer (1984) stated might be the case for much of public policy, the actual effect is to maintain the status quo. It also creates opportunities for profit.

In Sudan, food-based resilience practices such as medicalised nutrition, targeted food aid, and food vouchers were also convenient from a political and a logistical perspective. By 2008, much of the humanitarian operation in Darfur was managed remotely, with programmes implemented by national aid workers and/or organisations but managed by international staff. The Sudan government denied access to areas with rebel presence, and aid workers faced the risk of kidnap or attack (Stoddard et al. 2009). In such circumstances, it is easier to implement a standard package of interventions rather than develop context-specific approaches. Another advantage, from the perspective of international staff, is that risks are transferred to private sector operators or local NGOs. By 2010, moving food aid within Darfur had become almost impossible because of government access denials, militia presence, and numerous checkpoints operated by militia, rebel movements, and villagers.[3] In addition, with President Bashir's indictment for war crimes by the International Criminal Court, and the subsequent expulsion of 13 INGOs in 2009, gathering any information on the humanitarian situation in Darfur was extremely politically sensitive and risked expulsion. These issues are discussed further in the following sections, in particular how the practices of the resilience regime produced a depoliticised regime of truth in which food aid could be withdrawn amidst ongoing conflict and be presented as scientific progress.

Creating a fantasy: the function and dangers of regimes of untruths

In Darfur, food aid practices in the past 10–15 years have created a regime of truth in which conflict is invisible, people are malnourished because of their own actions, food security is minimal and where displaced people are lazy and dependent on food aid. This enabled international agencies to remain in Sudan, but it also facilitated Sudan's former government's counter-insurgency strategy; part of which was to control and manipulate food aid for political purposes. While the previous government has been ousted, the manipulation of food aid remains a feature (Jaspars and El-Tayeb 2021). Aid beneficiaries and long-term Sudanese aid workers understood the regimes of truth but contested them. This section first discusses what went missing in the practices of the resilience regime, then how food-based resilience practices led to a particular regime of truth, or fantasy, and finally discusses the functions this served.

What has been lost in the resilience regime?

Food-based resilience practices focus on nutrition and food security itself as the object of intervention, rather than its causes. They imply that by treating malnutrition with specialised food products, encouraging behaviour change, or implementing standardised packages of food security interventions, it is possible to create well-nourished and healthy people, that are able to adapt to or withstand shocks. As malnutrition also affects cognitive ability (Victora et al. 2008), better nutrition would also assist populations in making difficult decisions within an environment of severe resource constraints. Nutrition science is not alone in taking this cognitive turn. Since the World Bank's 2015 development report, entitled *Mind, Society and Behaviour*, cognitive science has become a key aspect of development, while ignoring its structural dimension (Duffield 2018). Furthermore, researchers and aid organisations have argued that nutrition interventions contribute to economic growth; in Sudan reportedly raising GDP by 3% (WFP and UNICEF 2014; Victora et al. 2008). However, focussing on treatment and behaviour change also means paying less attention to the structural social, political, and economic causes of malnutrition and food insecurity. The *Lancet* articles explicitly excluded these structural causes:

> Although addressing general deprivation and inequity would result in substantial reductions in undernutrition and should be a global priority, major reductions in undernutrition can also be made through programmatic health and nutrition interventions.
>
> (Black et al. 2008, 1)

> … excluded several important interventions which impact nutrition, such as education, untargeted economic strategies or those for poverty alleviation, agricultural modifications, farming subsidies, structural adjustments, social and political changes, and land reform.
>
> (Bhutta et al. 2008, 418)

This trend towards removing the social, political, and economic context from knowledge about food security and nutrition is further exacerbated by a trend towards quantitative assessments. From the early 2000s, WFP has been searching for a single quantitative indicator which can reflect the severity of food insecurity and which can be used to compare different areas or population groups (Coates et al. 2007). From the start of the Darfur crisis, WFP has used Dietary Diversity (DD) and the Food Consumption Score (FCS) as key indicators of food security. Yet, the use of this indicator has been little studied in emergencies and may not be suitable for food aid-dependent populations or for the ultra-poor (see, for example, Coates et al. 2007; Wiesmann et al. 2008). Furthermore, a number of studies find that these indicators tend to underestimate food insecurity compared to others (Maxwell et al. 2014; Leroy et al. 2015). So what could have been the reason for adopting the FCS as the key indicator of food insecurity? Assessments based on quantitative indicators say little about the nature or causes of food insecurity. Like the medicalisation of nutrition, it de-politicises. Ongoing violence by government-alligned militia, limitations in freedom of movement or access to land or employment, or the political nature of vulnerability is hidden in WFP's assessments. From an operational perspective, however, this has clear advantages. First, for aid organisations to work in Darfur, it was important that their assessments were unobjectionable and uncontroversial. Under the previous regime, reporting on the effects of conflict was likely to lead to expulsion. Second, given that much programming in Darfur is done remotely, quantitative data can be collected by mobile phone and analysed far from the crisis itself (Mock et al. 2016). Third, it delinks food security from nutrition; meaning that with high levels of acute malnutrition but relatively low levels of food insecurity, the belief that malnutrition is a result of behaviour or people's own actions can be more easily supported. Following the revolution, reporting on the causes of food insecurity and the effects of violence may be less contentious. On the other hand, ongoing violence in Darfur necessitates remote management and reinforces a focus on behaviour change.

While it can be argued that nutrition or food security assessments in the 1980s and 1990s never fully examined their political causes, a combination of qualitative and quantitative information provided for a context-specific analysis which incorporated the knowledge of Sudanese assessors and crisis-affected populations (Young and Jaspars 1995). In addition, one aspect of nutrition in emergencies in the 1990s, was to examine the constraints on aid reaching the most vulnerable. In the 2000s, evaluations have tended to focus on new practices such as food vouchers and recovery-type food interventions such as school feeding and food-for-work. Yet, food aid as general food distributions remains the major part of assistance in Sudan (Jaspars and El-Tayeb 2021). The last exercise to examine food aid targeting in Darfur was in 2008 (Young and Maxwell 2013).

Finally, resilience approaches have promoted the withdrawal of assistance:

> the 'resilience' objective is not to shift the burden of humanitarian response onto crisis victims' but that 'strengthening the resilience of households, groups and communities will enable them to enjoy greater autonomy and dignity and reduce the number of calls for short-term external assistance'.
>
> (UN OCHA 2014, 18)

As mentioned in the previous section, by the end of the 1990s, many people in Sub-Saharan Africa found themselves facing protracted crisis. Early warning had rarely led to early response, and targeting strategies rarely reached the most vulnerable (see for example (Buchanan-Smith and Davies 1995; Jaspars and Shoham 1999). Aid workers had to show ever higher levels of malnutrition to get a response, reflecting a 'normalisation of crisis' (Bradbury 1998). In the 2000s, aid practices have created a new regime of truth in which a general food distribution is no longer needed in response to high levels of acute malnutrition (WFP 2013b).

This regime of truth in which people are malnourished and food insecure can be maintained because information on the indicators of humanitarian crisis is limited and because of the distance between aid workers and crisis-affected populations. This is an emotional as well as a physical distance, which facilitates stereotyping and makes it easier to withdraw assistance. Even in conflict-affected Darfur, aid workers had adopted a stereotypical view that people did not know how to look after their children or that malnutrition was a cultural or behavioural issue:

> Mothers do now know how to cook.... They have a child every year which means they wean too early.
>
> (interview with aid worker)

> People in X have too many wives, and too many children... We distributed a ration for two months, ... [which] did not change anything.
>
> (interview with aid worker)

A regime of truth in which food aid is no longer needed also addresses concerns about food aid dependency in protracted crises. This has long been a concern of aid organisations and a reason for cutting aid when crises become protracted, regardless of the humanitarian situation and ongoing risks (see for example Macrae et al. 1997). Even the worst humanitarian failures can be seen as a success in that at least they did not create dependency (Keen 1994). In the almost 20 years of conflict in Darfur, discourse has changed from the need for international protection to one in which displaced populations are seen as lazy and cheats. Government officials and aid workers alike viewed the displaced as enjoying free goods and services in the camps, as having developed 'coping strategies' but at the same had little information about the actual risks that people continued to face or about humanitarian indicators such as malnutrition and mortality.[4]

Creating a fantasy

The regime of truth in which conflict-affected people are food insecure or malnourished because of their own actions can also be seen as a fantasy because it requires denial of permanent or repeated emergency. A fantasy involves denying or distorting reality (Marriage 2006, 489). The Darfur fantasy is characterised by aid workers being able to talk about food aid dependency despite food aid for

IDPs having steadily declined since 2008. It also means aiming for recovery interventions, when many villages are empty and access remains restricted. It means believing that acute malnutrition prevalences above the emergency threshold can be the result of poor feeding and hygiene practices while also knowing that access to land remains restricted and that attacks and displacement continue.

The fantasy created by aid practices in Darfur is similar to that discussed by Marriage (2006) in South Sudan. In South Sudan, aid workers created a fantasy about principled and sustainable programming by denying the effects of conflict and violence on being able to provide assistance. The limited impact of aid was instead attributed to aid dependency, lack of participation or non-compliance with humanitarian principles. This fantasy maintained funding and also provided psychological protection to aid workers. In Darfur, aid workers created a fantasy that food aid was no longer needed because malnutrition and food insecurity are the result of people's own actions. The creation of this fantasy was facilitated because of the difficulties of gathering information and it was necessary because providing material assistance had become almost impossible in the face of persistent access denial of ongoing violence. In the fantasy, not only was food aid no longer needed, but new practices of promoting resilience could be presented as scientific progress and thus an improvement on previous practices.

The Darfur fantasy includes several forms of denial (Cohen 2001). The first is interpretive denial: things are given a different meaning. A good example is how what were considered acceptable levels of acute malnutrition changed over time. Until the early 1990s, a prevalence of acute malnutrition above 15% was considered a crisis in need of emergency response, but by the late 1990s much higher levels were needed to elicit a response. By 2012, WFP no longer considered such a nutritional situation in need of general emergency food assistance (WFP 2013b). Perhaps a more significant form of denial, in the case of Darfur, is the denial which consists of simultaneously knowing and not knowing. Aid workers in Darfur, could talk at the same time about aid dependency, or malnutrition being a result of poor feeding practices, and about ongoing violence and displacement. They could also give examples of the harmful effects of reducing aid. One local aid worker, for example, explained the withdrawal of food aid after government relocation of displaced from camps to peri-urban areas as both a success because people were less dependent on aid but at the same leading to more crime because they had no other way of making a living. This kind of denial can be a way of dealing with the stress and moral dilemmas of working in humanitarian crises. Walkup (1997) gives detachment and reality distortion as one way that aid workers cope with being confronted with ongoing crisis and suffering that they are unable to address. Remote management and resilience practices provide excellent vehicles for this: international aid workers are already detached and resilience ideology provides a way of viewing limited aid distribution amidst ongoing crisis as positive and promoting resilience. Another form of denial is that aid workers simply stop investigating causes they cannot address. Field staff in Darfur, for example, in response to a nutritional survey examining differences in households with malnourished and well-nourished children preferred to focus on the ability of mothers to care for their children rather than the differences in access to land or employment (Ibrahim 2011). Similarly, national nutrition surveys examine mostly health, sanitation, and behavioural indicators in relation to nutritional status, and thus conclude that the solution is treatment with RUTF and education on hygiene and feeding behaviours (Federal Ministry of Health 2014). While at some level, aid workers know that conflict and violence are ongoing, both the official and the day-to-day narrative are that malnutrition and food insecurity are due to people's own actions, and that resilience can be achieved by changing their behaviour. This fantasy builds up over time and provides psychological protection for aid workers. It also has a number of political and economic functions.

The risks and functions of regimes of alternative truths

The fantasy allows aid agencies to remain in Darfur, but it also fails to examine what food aid practices actually do in Sudan, which includes supporting a private sector which is closely alligned to the Sudan government, facilitating counter-insurgency operations, as well as making conflict and power relations invisible (Jaspars 2018b). From the first food aid operations in the 1980s, transporters and traders have benefited from food aid. Until the 1990s, this often involved delaying operations to increase the cost and maximise profits, and in the early 2000s, a limited number of transporters grew into multi-national operations as a result of WFP contracts (ibid.). The reduction of food aid for long-term IDPs from 2008, converged with government counter-insurgency tactics and policies of emptying the camps. The government developed a policy of encouraging return in 2010 (Government of Sudan 2010) and the previous government used increasingly forceful strategies to bring this about. The former government's denial of access to rebel-held areas meant that by 2014, international food aid mostly went to those living in government-controlled areas. By 2017, most of Darfur was in government hands, and through food or financial incentives many of the leaders in the IDP camps had been brought over to the government side (Jaspars and Buchanan Smith 2018). Camp leadership has changed since the 2019 revolution, but food aid remains highly targeted and return is key aspect of the peace agreement between the Transitional Government and some of the rebel movements (Jaspars and El-Tayeb 2021).

A fantasy in which malnutrition and food insecurity is a result of cultural and behaviour factors also removes government responsibility for creating food insecurity through its war strategies, denial of access, and manipulation of humanitarian assistance. It removes international responsibility for protecting civilians from large-scale loss of life at the hands of their own government. Viewing malnutrition as the result of individual behaviour, as discussed earlier, shifts responsibility from the state or international actors to the individual.

This contrasts with the social nutrition of the 1980s and 1990s, which treated malnutrition as a social and political problem at the level of populations (Jaspars

2019). The disappearance of politics, or the government's war strategies, was also convenient for the EU because it enabled collaboration with the Sudanese government on stemming migration to Europe. Few projects are implemented in Darfur, and they are largely focussed on livelihoods, food security, and resilience (Jaspars and Buchanan Smith 2018). This too is based on fantasies, first that food security will stop migration, second that people migrate because they are not resilient and conversely that resilience means staying in one place. This ignores the wealth of literature on migration, on famine and on the importance of mobility as a livelihood strategy (Carling and Talleraas 2016; Sadliwala 2019; Young and Ismail 2019).

Finally, a resilience regime consisting of medicalised nutrition, quantified food security, and behaviour change, suppresses the views, perceptions, and experiences of conflict-affected people, and Sudanese aid workers. For them, both international and government food aid is used as a political tool but at the same time continues to be needed for people who continue to face threats to their livelihoods and difficulties in accessing food.[5] With the dominant aid agency and government regimes, however, and the remoteness of international organisations, the reality of local populations remains separate and with apparently little formal influence over food aid practices.

Resilience, vulnerability, or abandonment?

For most people in Darfur, the resilience regime has been associated with a reduction in food aid, increasing food prices, and ongoing violence and threats to livelihoods. From 2008 food aid (or food assistance) for many conflict-affected has decreased, in part because of funding, declines in access (at first insecurity later denial of access by the previous government) but also an assumption that people could meet part of their own food needs (Jaspars 2018a). By 2017, only 40% of IDPs in camps received food assistance to meet only a small portion of their food needs, and rural populations only when they had experienced a shock like drought or floods. Part of the aim was to reduce dependency (WFP 2015). In 2021, a similar system remained (Jaspars and El-Tayeb 2021). The previous section showed that this reduction in food aid facilitated counter-insurgency and policies to empty the IDP camp and/or to bring their leaders over to the government side under the previous regime.

As food aid decreased and international organisations adopted resilience practices, the former government and its aligned militia continued to attack and cause death and destruction in Darfur. From 2013 to 2016, violent conflict and displacement increased. This was to a large extent associated with the creation of the Rapid Support Forces (RSF), a paramilitary group formed from militia which fought alongside the government. RSF attacks, conflict between Arab militia and between militia and government, caused the displacement of over 1.2 million people between 2013 and 2016 (UN OCHA 2018). Confrontation between the Sudan Armed Forces and the rebel movements halted temporarily in 2017, but resumed in 2018 resulting in renewed displacement from Jebel Marra, which had been a rebel stronghold. At the same time, militia attacks on IDP camps and rural populations continued with impunity (UN OCHA 2019). In 2021, despite a change in government and peace efforts in Darfur, violence and displacement increased. The number in need of humanitarian assistance in Sudan at the time of writing was on an unprecedented level (Jaspars and El-Tayeb 2021; UN OCHA 2021a).

While information is limited, indications are that most people's livelihoods remain extremely precarious, dependent on marginal activities such as casual labour, selling firewood and charcoal or petty trading (WFP 2020). In 2017, displaced populations continued to face restrictions in movement due to risk of attack and young men were frequently detained on suspicion of rebel activities (Jaspars and Buchanan Smith 2018). Their original land, for many, remains occupied by Arab nomadic groups whose fear of losing out in the peace agreement is now manipulated by elements of the former regime (Jaspars and El-Tayeb 2021). Criminal or militarised strategies increased throughout the conflict; including, for example, demanding protection payments, joining militia or paramilitary groups, as well as looting and theft. Between 2014 and 2016, an increasing number of Darfuris (in particular the ethnic groups associated with the rebellion) migrated to Europe because of the ongoing risks to their safety and the limited livelihoods options in Darfur (Jaspars and Buchanan Smith 2018).

In Darfur, resilience approaches leave people to try and find any means to survive in a permanent emergency. What is also clear, is that the survival of some groups is dependent on exploiting and harming others. It appears to be everyone's individual responsibility to keep safe, find food and prevent malnutrition. As Evans and Reid (2013) write, one the one hand, it makes sense to support people taking on these responsibilities, but on the other hand it has a dehumanising political effect. It fails to take account of the power of political resistance against the conditions that cause suffering. A key question is therefore what resilience actually is and what resilience approaches do in Darfur. Do the precarious, criminal, or desperate migration strategies represent resilience? Is it survival while remaining vulnerable to attack, harassment and other forms of persecution? It appears that in this case, the practices of the resilience regime actually maintain the vulnerability of segments of the population, because they ignore power relations or the structural causes of conflict and crisis in Darfur. Rather than addressing its causes, they expect crisis-affected populations themselves to find solutions. As such it also represents an abandonment of crisis-affected populations.

Conclusions

In Darfur, resilience practices have led to an abandonment of crisis-affected populations. Resilience practices can be seen as a form of neoliberal governmentality because they urge the creation of responsible subjects who can adapt to permanent emergency. By focussing on treatment and behaviour change to improve food security, its structural causes are not addressed and conflict and power relations have been made invisible. Generally accepted emergency levels of acute malnutrition, are no longer considered to need a general food aid or food assistance response. Compared to the 1980s and 90s, the term 'vulnerability' has been used to further target and reduce material assistance. This in turn facilitated the previous government's counter-insurgency which included the manipulation of food aid. For Western nations, making conflict invisible was convenient as it allowed for collaboration with Sudan's previous regime on stopping migration to Europe and enabled collaboration on trade and business. It remains to be seen whether the new transitional government can bring peace and prevent human rights abuses in Darfur.

In Sudan, resilience has become a smokescreen for the ongoing persecution and human rights violations in Darfur. Rather than engaging with the Sudan government on human rights abuses as causes of migration, health and nutrition and agricultural support can give the illusion of resilience and migration management in Darfur. Cavelty et al. (2015) suggested that resilience is a chimera, a vision of something that relates to past events, and ideally to the future, but never actually in the present. Howell (2015) asserts that ideas of resilience-oriented governance are to some extent fantastical; as failure (or contestation) is inherent in governance. This chapter supports these conclusions and takes them further. In Darfur, resilience is a dangerous fantasy – that has been produced through practices that place responsibility for survival on the individual while conflict and crisis are ongoing. They have contributed to the normalisation and human rights abuses in Sudan.

Compared to aid critiques of the past, challenging resilience practices is difficult. In Keen's (1994) work on the political economy of famine and relief in Sudan, for example, the failure of relief in reaching crisis-affected population was generally acknowledged as a failure and his work added a new dimension in that it revealed that it was a success for government, merchants and those with commercial farms. In contrast, international organisations view the limited assistance provided in the resilience regime as a success, and as a way of working in protracted crisis. However, while resilience practices may give the illusion of success for aid workers, private sector, government, and donors, it is not a success for crisis-affected populations themselves.

The chapter has also raised key questions about the role of crisis-affected populations themselves and the role of the private sector. How can ordinary people resist or influence the diffuse power of aid practices in the resilience regime? Can they bring about a change in the system, rather than simply adapt their behaviour or ignore, co-opt or adapt international aid practices? Opportunities for resistance appear limited in the face of dominant international and government practices, and people's actual responses – like migration to Europe – are not an approved form of resilience and instead considered a crime. To give a more complete picture of what resilience practices are actually doing, further research is needed on people's ability to resist the governmental effects of aid, including an examination of the nature and extent of private sector involvement, and the role of resilience practices in containment.

Notes

1 This is a shorter and updated version of an article published on 13 July 2020 in *Security Dialogue*. https://doi.org/10.1177%2F0967010620927279. The chapter came about through funding from the European Research Council (ERC) under the European Union's Horizon 2020 research and innovation programme, grant agreement No 884139.
2 Interview with USAID representatives, 2014.
3 Interviews with aid agency staff, 2013 and 2014.
4 Interviews in Khartoum and El Fasher in 2012 and 2013.
5 Interviews in El-Fasher, September–October 2013.

References

Arnold, Tom and David Beckmann. 2011. "Update on Scaling up Nutrition (SUN) and the '1000 Day' movements." In *Field Exchange* 41, edited by Asfaw, A. and Ibrahim, S., Oxford: Emergency Nutrition Network. https://www.ennonline.net/fex/41/en.

BBC (2004) "Mass rape atrocity in west Sudan," Accessed March 9, 2014. http://news.bbc.co.uk/1/hi/world/africa/3549325.stm.

Bhutta, Zulfiqar, Tahmeed Ahmed, Robert Black, Simon Cousens, Kathryn Dewey, Elsa Giugliani, Batool Haider, et al. 2008. "Maternal and Child Undernutrition 3: What works? Interventions for maternal and child undernutrition and survival." *Lancet* 371: 417–440.

Black, Robert, Lindsay Allen, Zulfiqar Bhutta, Laura Caulfield, Mercedes de Onis, Majid Ezzati, Colin Mathers, and Juan Rivera. 2008. "Maternal and Child Undernutrition 1. Maternal and child undernutrition: Global and regional exposures and health consequences." *Lancet* 371: 243–260.

Bradbury, Mark. 1998. "Normalising the crisis in Africa." *Disasters* 22 no. 4: 328–338.

Buchanan-Smith, Margie. 1989. *Evaluation of the Western Relief Operation 1987/88. Final report.* Wokingham: MASDAR (UK) Ltd.

Buchanan-Smith, Margie and Susanna Davies. 1995. *Famine Early Warning and Response - The Missing Link.* London: Intermediate Technology Publications.

Buckley, Ruth. 1988. "Food targeting in Darfur: Save the children fund's programme in 1986." *Disasters* 12 no. 2: 97–103.

Carling, Jorgen and Cathrine Talleraas. 2016. *Root Causes and Drivers of Migration: Implications for Humanitarian Efforts and Development Cooperation.* Oslo: Peace Research Institute Oslo.

Cavelty, M., Kaufmann, M. and Kristensen, K. 2015. "Resilience and (in)security: Practices, subjects, temporalities." *Security Dialogue* 46 no. 1: 3–14.

Coates, Jennifer, Beatrice Rogers, Patrick Webb, Daniel Maxwell, Robert Houser, and Christine McDonald. 2007. *Diet Diversity Study.* ODAN. Emergency Needs Assessment Service. Rome: World Food Programme.

Cohen, Stanley. 2001. *States of Denial: Knowing about Atrocities and Suffering.* Cambridge: Polity.

DfID. 1997. *Report on Sudan Emergency Food Distributions.* London: Department for International Development.

Duffield, Mark. 2016. "The resilience of the ruins: Towards a critique of digital humanitarianism." *Resilience* 4 no 3: 147–165.

Duffield, Mark. 2018. *Post-Humanitarianism: Governing Precarity in the Digital World.* Cambridge: Polity Press.

European Commission. 2016. *Actions in Support of Tackling Irregular Migration and Displacement in the Horn of Africa within the EU Emergency Trust Fund*. Brussels: EU.

Evans, Brad and Julian Reid. 2013. "Dangerously exposed: The life & death of the resilient subject." *Resilience: International Policies, Practices & Discourses* 1 no: 2: 83–98.

FAO. 2008. *High-Level Task Force on the Global Food Crisis: Comprehensive Framework for Action*. Rome: Food and Agriculture Organisation.

FAO. 2012. *High Level Expert Forum on Addressing Food Insecurity in Protracted Crisis: Report*. Rome: Food and Agricultural Organisation.

Federal Ministry of Health. 2014. *Report of a Simple Spatial Surveying Method (S3M) Survey in Sudan*. Khartoum: Federal Ministry of Health.

Fewsnet. 2021. *SUDAN Food Security Outlook. February to September 2021. Conflict and a Macroeconomic Crisis Drive Above Above-Average Needs Through September 2021*. Khartoum: USAID.

Foucault, Michel. 1980. "Truth and power." In *Power/Knowledge: Selected Interviews and Other Writings*, edited by Colin Gordon, pp. 109–133. New York: Pantheon Books Translated by Colin Gordon, Leo Marshall, John Mepham, Kate Soper.

Foucault, Michel. 2007. *Security, Territory, Population. Lectures at the College de France 1977–78*,. Basingstoke: Palgrave. Translated by Graham Burchell.

Global Compact for Migration. 2018. *Global Compact for Safe, Orderly and Regular Migration*. Morocco: Global Compact for Migration.

Government of Sudan. 2010. *Darfur: Towards New Strategy to Achieve Comprehensive Peace, Security and Development*. Khartoum: Government of Sudan.

Hilhorst, Dorothea. 2018. "Classical humanitarianism and resilience humanitarianism: Making sense of two brands of humanitarian action." *Journal of International Humanitarian Action* 3 no 15. https://jhumanitarianaction.springeropen.com/articles/10.1186/s41018-018-0043-6#citeas

Howell, Alison. 2015. "Resilience as enhancement: Governmentality and political economy beyond responsibilisation." *Politics* 35 no. 1: 67–71.

Ibrahim, Insaf. 2011. *Study Report: Causality Study on Causes of Persistent Acute Malnutrition in North Darfur (Kabkabyia). 2010–2011*. Khartoum: Word Food Programme.

Ilcan, Suzan and Kim Rygiel. 2015. ""Resiliency Humanitarianism": Responsibilising refugees through humanitarian emergency governance in the camp." *International Political Sociology* 9: 333–351.

Jaspars, Susanne. 2018a. *Food Aid in Sudan. A History of Power, Politics and Profit*. London: Zed Books.

Jaspars, Susanne. 2018b. "The state, inequality, and the political economy of long-term food aid in Sudan." *African Affairs* 117 no. 469: 592–612.

Jaspars, Susanne. 2019. *A Role for Social Nutrition in Strengthening Accountability for Mass Starvation?*. World Peace Foundation Occassional Paper 20. Boston: Tufts University.

Jaspars, Susanne and Margie Buchanan Smith. 2018. *Darfuri Migration from Sudan to Europe: From Displacement to Despair*. Joint study by REF (SOAS) and HPG. London: Overseas Development Institute.

Jaspars, Susanne and Youssif El-Tayeb. 2021. *Caught in Transition: Food aid in Sudan's Changing Political Economy*. Khartoum: Saferworld Conflict Sensitivity Facility.

Jaspars, Susanne and Jeremy Shoham. 1999. "Targeting the vulnerable; the necessity and feasibility of targeting vulnerable households." *Disasters* 23 no. 4: 359–372.

Joseph, Jonathan. 2016 "Governing through failure and denial: The new resilience Agenda." *Millennium: Journal of International Studies* 44 no. 3: 370–390.

Karim, Ataul, Mark Duffield, Susanne Jaspars, Antonio Benini, Joanna Macrae, Mark Bradbury, Douglas Johnson, George Larbi, and Barbara Hendrie (1996) *Operation Lifeline Sudan: A Review.* Birmingham: University of Birmingham.

Keen, David. 1994. *The Benefits of Famine: A Political Economy of Famine and Relief in South Western Sudan 1983–89.* Oxford: James Curry.

Leroy, Jef, Marie Ruel, Edward Frongillo, Jody Harris, and Terri Ballard. 2015. "Measuring the food access dimension of food security: A critical review and mapping of indicators." *Food and Nutrition Bulletin* 36 no. 2: 167–195.

Levine, Simon and Irina Mosel. 2014. *Supporting Resilience in Difficult Places. A critical look at applying the 'resilience' concept in countries where crises are the norm.* HPG Commissioned Report. London: Overseas Development Institute.

Macrae, Joanna, Mark Bradbury, Susanne Jaspars, Douglas Johnson, and Mark Duffield. 1997. "Conflict, the continuum and chronic emergencies: A critical analysis of the scope for linking relief, rehabilitation and development planning in Sudan." *Disasters* 21 no 3: 223–243.

Marriage, Zoe. 2006. "The comfort of denial: External assistance in Southern Sudan." *Development and Change* 37 no 3: 479–500.

Maxwell, Daniel, Bapu Vaitla, and Jennifer Coates. 2014. "How do indicators of household food insecurity measure up? An empirical comparison from Ethiopia." *Food Policy,* 47: 107–116.

Mock, Nancy, Gaurav Singhal, William Olander, Jean-Baptiste Pasquier, and Nathan Morrow. 2016. "mVAM: A new contribution to the information ecology of humanitarian work." *Procedia Engineering* 159: 217–221.

Ockwell, Ron. 1999. *Thematic Study on Recurring Challenges in the Provision of Food Assistance in Complex Emergencies.* Rome: World Food Programme.

Osman, El-Fateh. 1993. *Targeting Realities of Food Aid: The Experience of Darfur State: SCF Food Aid Meeting.* Nairobi: Save the Children-UK.

Sadliwala, Batul. 2019. "Fleeing mass starvation: What we (don't) know about the famine-migration nexus." *Disasters.* doi:10.1111/disa.12420.

Schaffer, Bernard. 1984. "Towards Responsibility: Public Policy in Concept and Practice." In *Room for Manouvre: An Exploration of Public Policy in Agriculture and Rural Development,* edited by Edward Clay and Bernard Schaffer, pp. 142–190. London: Heinemann Educational Books.

Scott-Smith, Tom. 2018. "Paradoxes of resilience: A review of the world disasters report 2016." *Development and Change* 49 no 2: 662–677.

Stoddard, Abby, Adele Harmer, and Victoria DiDomenico. 2009. *Providing aid in Insecure Environments: 2009 Update.* HPG Policy Brief 34. London: Overseas Development Institute.

Street, Alice. 2015. "Food as pharma: Marketing nutraceuticals to India's rural poor." *Critical Public Health* 25 no 3: 361–372.

SUN (2014) *An Introduction to the Scaling Up Nutrition Movement.* Geneva: Scaling Up Nutrition Movement.

UN OCHA. 2014. *2014 Revised Strategic Response Plan.* Khartoum: UN Office for the Coordination of Humanitarian Affairs.

UN OCHA. 2018. *2018 Humanitarian Needs Overview Sudan.* Khartoum: UN OCHA.

UN OCHA. 2019. *Humanitarian Bulletin Sudan. Issue 01|24 December 2018–27 January 2019.* Khartoum: UN OCHA.

UN OCHA. 2021a. *Humanitarian Needs Overview Sudan.* Khartoum: UN OCHA.

UN OCHA. 2021b. *Humanitarian Response Plan: Sudan: 2021*. Khartoum: UN OCHA.

UNHCR. 2018. *Report of the United Nations High Commissioner for Refugees. Part II. Global Compact on Refugees*. General Assembly Official Records Seventy-third Session Supplement No. 12. Geneva: UNHCR.

UNICEF. 2013. *Bridging the Nutrition Security Gap in Sub-Saharan Africa: A Pathway to Resilience and Development*. Paris: United Nations Children's Fund.

Victora, Cesar, Linda Adair, Caroline Fall, Pedro Hallal, Reynaldo Martorell, Linda Richter, and Harshpall Singh Sachdev. 2008. "Maternal and Child Undernutrition 2. Maternal and child undernutrition: Consequences for adult health and human capital." *Lancet* 371: 340–357.

Walkup, Mark. 1997. "Policy dysfunction in humanitarian organizations: The role of coping strategies, institutions, and organizational culture." *Journal of Refugee Studies* 10 no 1: 37–60.

Welsh, Marc. 2014. "Resilience and responsibility: Governing uncertainty in a complex world." *The Geographical Journal* 180 no 1: 15–26.

WFP. 2013a. *WFP Strategic Plan (2014–2017)*. Executive Board Session. 3–6 June Rome: World Food Programme.

WFP. 2013b. *Sudan: An Evaluation of WFP's Portfolio 2010–2012. Vol I - Evaluation Report*. Country Portfolio Evaluation. Rome: World Food Programme.

WFP. 2015. *Policy on Building Resilience for Food Security and Nutrition*. Executive Board Annual Session Rome, 25–28 May. Rome: World Food Programme

WFP. 2017. *WFP Strategic Plan (2017–2021)*. Rome: World Food Programme.

WFP and UNICEF. 2014. *The Case for Investment in Nutrition in Sudan*. Khartoum: WFP and UNICEF.

WFP Sudan. 2020. *Food Security Monitoring System (FSMS) Sudan. Q1 2020*. Khartoum: Word Food Programme.

Wiesmann, Doris, Lucy Bassett, Todd Benson, and John Hoddinott. 2008. *Validation of Food Frequency and Dietary Diversity as Proxy Indicators of Household Food Security Report. Submitted to: World Food Programme*. Rome: World Food Programme.

Young, Helen, Annelies Borrel, Dianne Holland, and Peter Salama. 2004. "Public nutrition in complex emergencies." *The Lancet* 364: 1899–1909.

Young, Helen and Musa Ismail. 2019. "Complexity, continuity and change: Livelihood resilience in the Darfur region of Sudan." *Disasters* 43 no. S3: S318–S344.

Young, Helen and Susanne Jaspars. 1995. *Nutrition Matters: People, Food and Famine*. Rugby: Intermediate Technology Publications.

Young, Helen and Daniel Maxwell. 2013. "Participation, political economy and protection: Food aid governance in Darfur, Sudan." *Disasters* 37 no 4: 555–578.

Young, Helen, et al. 2009. *Livelihoods, Power and Choice: The Vulnerability of the Northern Rizeigat, Darfur, Sudan*. Boston: Feinstein International Center. Tufts University.

9

VULNERABILITY AND RESILIENCE IN A COMPLEX AND CHAOTIC CONTEXT

Evidence from Mozambique

Luís Artur

Climate Change (CC) and related disasters represent a serious threat to human existence. They particularly affect the poorest people living in the most vulnerable countries and communities. A World Bank report (World Bank 2017) states that the impact of climate change and related disasters on the poor and extremely vulnerable are 60 per cent higher compared to other social groups. This is related, amongst other factors, to the fact that poor and vulnerable people work in sectors highly susceptible to climate conditions such as in agriculture, and live in areas and houses highly exposed and vulnerable to extreme events. Hence, climate change adaptation (CCA) and disaster risk reduction (DRR) programmes should preferably aim to target these groups of people if the UN's Sustainable Development Goals, the Sendai Framework, and aid harmonisation are to be achieved.

Poor and vulnerable people increasingly find themselves at the heart of national and international social protection (SP) mechanisms. Social protection is commonly understood as all public and private initiatives that provide income or consumption transfers to the poor, protect the vulnerable against livelihood risks, and enhance the social status and rights of the marginalised with the overall objective of reducing the economic and social vulnerability of poor, vulnerable and marginalised groups (Devereux and Sabates-Wheeler 2004, i). It is a novel field initiated in the early 2000s with the publication in 2003 of the first World Social Protection Report by the World Bank (World Bank 2003).

SP has now become a fast-growing global policy sector, tripling in real expenditure over the last two decades (Simson 2020). Due to its role and global interest, it has been argued that SP should be aligned with CCA and DRR interventions. Notwithstanding, SP mechanisms are, in most cases, detached from interventions on adaptation and DRR reducing the overall impact of social programmes on vulnerability reduction. A new concept, Adaptive Social Protection or Shock

DOI: 10.4324/9781003219453-11

Responsive Social Adaptation, that links DRR and CCA to SP, is emerging but the methodology for such a new concept still requires framing and is the scope of this chapter.

This chapter, based on a case study in Mabote district in Mozambique, argues that CCA, DRR and SP interventions risk failing to lift people out of poverty and vulnerability because they use inappropriate targeting tools and interventions. Current tools and interventions are still mainly top-down and biased towards an inappropriate concept of resilience. Resilience, which refers to the ability of a social and ecological system to resist, absorb, adapt, and maintain itself in face of external threats (Adger 2000) must be treated with some caution. The scope for resilience among people (such as those who are part of the social protection programmes) whose capabilities have been squeezed or eroded by poverty, climate change, and the demands for flexibility in a competitive market model of development may be questionable (Hadded et al. 2011, 5). On the other hand, as emphasised by McGregor (2011), the resilience of one part of a system is often achieved at the cost of a greater burden of work and self-exploitation by women, children, the elderly, and people with disabilities. The resilience of firms, too, is often dependent upon flexible, poorly paid and exploitative labour markets, while governments, at global and national levels, depend for their macroeconomic resilience on cuts to public investments to services such as social protection and asking the poor and vulnerable to be more resilient (McCulloch and Grover 2010).

The concept of resilience tends to mask the growing local and international inequality which produces vulnerability and poverty. The World Inequality Report (WIR 2018) shows that, at the global level, inequality has risen sharply since 1980. From 1980 to 2016, the global top one per cent of earners has captured twice as much of growth as the 50 per cent of poorest individuals. Across the world, the gap between rich and poor is rapidly increasing. Irish (2017, 5) argues that almost half of the world's wealth is now owned by just one per cent of the population, and seven out of ten people live in countries where economic inequality has increased in the last 30 years. The World Economic Forum Global Risk Report (WEF 2017) identifies economic inequality as a major risk to human progress, impacting social stability within countries and threatening security on a global scale. Findings by different scholars such as Berg and Ostry (2011), Ravallion (2005) suggest that chronic inequality stunts long-term economic growth and makes it more difficult to reduce poverty. Overall, the different reports suggest that institutions and policies matter for tackling inequality and poverty.

In Mozambique, Gradin and Tarp (2017) show that the growth pattern that led to a reduction in poverty over time went along with a substantial increase in inequality, especially in most recent years. Arndt and Mahrt (2017) conclude that inequality is worse than previously reported because the consumption of the better offs is under-reported; wealthier households are able to consume three times what the bottom 40 per cent consume (World Bank 2018). The World Bank report concludes that, although the poor gained from growth, the wealthier segments of society gained more and at a faster pace, especially in urban areas. This recent 'pro-richness' pattern reflects the extent to which Mozambique's growth acceleration and improvements in access to services have been concentrated in urban centres, hindering the nation's progress in achieving shared prosperity and making it among the most unequal countries in sub-Saharan Africa today. Finally, a recent financial scandal in Mozambique, in which USD $2 billion 'disappeared' only to enrich local elites, reveals how policies and institutions matter for reimagining development.[1] It also stresses the need to look at resilience beyond individual capabilities as many authors tend to do when analysing resilience. As ably put by Krugman (2009), 'don't mistake truth for beauty'.

Based on evidence from Mozambique, this chapter attempts to fuel the discussion on whether adaptive social protection can promote resilience. It commences with a discussion of vulnerability and resilience in today's world before introducing a case study on ASP to explore its existing limitations to promote resilience.

Vulnerability and resilience in a complex, chaotic and (un) globalised world

At the writing of this chapter (end of 2020), Mozambique is still attempting to recover from two major cyclones (Idai and Kenneth in March and April 2019) which together affected more than 1.5 million people and claimed more than 600 lives. Thousands of classrooms and hectares of cropped land were washed away, and many thousands of people are still dependent on humanitarian assistance to survive. The national government pledged about USD $3.2 billion for reconstruction. The country also received promises of aid to the amount of USD $1.02 billion of which only USD $294.92 million had been disbursed by the end of 2019 with a further USD $144.17 million expected to be injected by 2020 (GREPOC 2020). Due mainly to the previously discussed financial scandal, donors have tightened good governance measures and are demanding transparency, clear bankable projects, and a preference to allocate funds through international agencies. Mozambique is also plagued by terrorist activity in the northern province of Cabo Delgado that has already displaced about 500,000 people and unleashed a military counter insurgency in the central region. The economy has also been hit hard by COVID-19 impacts.[2]

Apart from the hazard (Cyclone Idai and average seasonal rainfall), the 2019 flooding in Mozambique took place in a complex and chaotic context. Rainfall upstream in Malawi and Zambia led to a collapse of dams and bridges in those countries but mainly downstream in the floodplains of Mozambique. By March 12, nearly 50,000 people had been affected in Malawi, resulting in 30 recorded deaths and over 370 injured (Reliefweb 2019). In Mozambique, the city of Tete was partly submerged due mainly to dam and bridge collapses in Malawi. This event happened in a context where illegal construction of houses, including in the vicinity of bank rivers, has become normal practice. At the same time, the country's economy was under considerable strain due to the already mentioned financial crisis that led to the withdraw of major donors – Mozambique depends for about 40 per cent of its annual budget on international donors (Banco Mundial 2016). There was also a drastic reduction in public and private investments across the country. According to the African Development Bank (2019), Mozambican real GDP growth was estimated to be 3.5 per cent in 2018, a dramatic decline from the average seven per cent achieved between 2004 and 2015. This decline was due to decreased public investment and a 23 per cent decrease in foreign direct investment 2015–2017. The fiscal deficit was an estimated 6.7 per cent of GDP in 2018, up from 5.5 per cent in 2017. Since the discovery of the hidden debt in 2016, Mozambique has been in default with the result that major donors suspended aid to the country. Consequently, it has been forced to implement fiscal measures to gradually reduce this public debt (Banco Mundial 2016; Francisco and Semedo 2016; Francisco and Semedo 2017; Aiuba 2019). To achieve these cuts in public expenditure, public services such as agricultural extension and road and bridge maintenance amongst others have been sacrificed. Rising international prices for key imports such as fuel and food have only exacerbated an already fragile context.

During 2016–2017, Mozambique had high inflation and a rapidly depreciating exchange rate which forced the Bank of Mozambique to ease monetary policy, lowering the benchmark lending rate to 18 per cent in August 2018. However, the decrease in inflation from 15.1 per cent in 2017 to an estimated 4.6 per cent in 2018 was mainly due to the reduction in people's purchasing power rather than an increase in supply (Nova 2019). The situation was also fuelled by a contraction in credit demand by the private sector (African Development Bank 2019). A study by Feijo (2018) found that 76.7 per cent of his respondents across the country experienced at least one month of food shortage in 2016, and the increasing number of destitute people was becoming a new pool of cheap labour for the emerging elite in rural areas. Nova (2019) found that in 2018 poor people had very low purchasing power, so increasing their likelihood for malnutrition and poverty.

Beyond these domestic considerations, the global situation has become less friendly for migration and globalisation. National government is progressively putting more restriction on migration and international trade. The rise in nationalist movements and conservative leaders across the world is further increasing vulnerabilities worldwide. South Africa – the 'el dorado' for Africans, especially Mozambicans – is in the throes of xenophobia and has tightened its migration policies, squeezing many Mozambicans who traditionally and (in) formally depended on working there for their livelihoods. Media images of a Mozambican, Ernesto Nhamuave, burned alive by South Africans and of another Mozambican, Emídio Macie, being brutally beaten, tied, and dragged in a police car caught international attention and raised serious doubts about the future of globalisation as well as the economic and social resilience of societies. Much the same is valid in Europe where massive migration movements have led to nationalist movements and, to some extent, a rise in xenophobia. International trade has also suffered. Governments worldwide, especially in high-income countries, have increased the use of non-tariff barriers such as licensing requirements, border controls, certificates and so on to an extent that poor countries find it hard to trade. For instance, under the AGOA (African Growth and Opportunity Act), a US government programme to boost the sale of goods produced in Africa, Mozambique can export 6,500 different products to the American market but has taken little advantage of this opportunity. In 2016 when the US African Export and Administration Policy coordinator, Florizelle Liser, visited Mozambique, spokespersons from the private sector complained about the excessive barriers to US markets. Paulo Fumane, the Mozambican private sector chief adviser at the time, went on to say:

> In recent years, some sectors have reduced or almost stopped exporting to the American market. It is necessary to reactivate the trade relations by seeking ways to remove the barriers, which are not tariffs but rather technical requirements in export chains such as on licensing, sanitary requirements to mention a few
>
> (Club of Mozambique 2016)

Yalcin et al.'s (2017) study on trade concludes that globally the number of non-tariff barriers have jumped from 389 in 2009 to 2,421 in 2016. The US, with 800 non-tariff barriers implemented, was number one, followed by India 310 and Russia 240. Larger European economies such as Germany, the UK, and France come next, implementing between 50 and 100 non-tariffs barriers over the study period. The Global Risk Report (WEF 2019) summarises the present world situation in a very eloquent way:

> Global risks are intensifying but the collective will to tackle them appears to be lacking. Instead, divisions are hardening. The world's move into a new phase of strongly state-centred politics… The idea of 'taking back control' – whether domestically from political rivals or externally from multilateral or supranational organizations – resonates across many countries and many issues. The energy now expended on consolidating or recovering national control risks weakening collective responses to emerging global challenges. We are drifting deeper into global problems from which we will struggle to extricate ourselves.
>
> (WEF 2019, 5)

Under such environmental, economic, political, and financial circumstances, we need to redefine and reimagine the concept of resilience. As previously mentioned, the scope for resilience among people whose capabilities have been squeezed or eroded by a context such as pictured above may be questionable. Instead, what we are experiencing makes it clear how vulnerability is being produced, reproduced, and/or expanded and, therefore, why it still matters. The pressure and release model by Wisner et al. (2004), which shows that vulnerability has its roots in macro (international and national) policies of exclusion and inequality, applies equally to our ongoing global deficits.

It is also time to realise that narrow and compartmentalised schools of thought no longer have the authority to depict present-day complexities. Rather, we need a shift towards ecological science, chaos and complexity theories, and systemic thinking in order to provide the right answers to the current crisis. As argued by Burns (2011, 35), in a context of global crisis like the one we are living through, it is not enough to look at particular aspects or cases; they are intrinsically connected to the other elements of the system which surround them. 'We need to use multiple lenses, and from multiple perspectives to explore where new visions of development, *vulnerability and resilience* might lie' (emphasis added).

Social protection: institutionalising the excluded

Social protection is a field in the making. Over the past 10 years, social protection has received substantial and increasing academic, economic, and political attention from a diverse range of stakeholders. Evidence shows that investments in appropriate and well designed and delivered social protection systems and programmes can lead to substantial positive outcomes. These range from improvements in measures of household wellbeing (in terms of income, assets, dignity, and food security), to better access to services, better nourished and educated children, reduced vulnerability and improved resilience to shocks, and more sustained and sustainable local and national economic growth. Cash and food transfers are also associated with a decline in early marriage and pregnancy, recourse to transactional sex and HIV infection among young women and girls. In sum, it is advocated that social protection works for the poor, and is a central element of policy frameworks for more inclusive and equitable economic growth (Irish 2017). As such and under growing pressure due to inequalities, climate change, and market distortions, social protection policies and strategies are being taken up by many governments and development partners. Organisations as prominent as UNICEF, the World Bank, the European Union, G20, the African Union, and the International Labour Organisation (ILO) have adopted policies and strategies that not only highlight the role of social protection in economic growth and economic empowerment but ones that also emphasise the importance of social protection for addressing human, social, and political rights.

The World Bank (2011) suggests that social protection plays three critical roles for the 'beneficiaries'– protect–prevent–promote (the 3Ps):

- Protective role: providing relief from deprivation (e.g. income benefits, state pensions)
- Preventative role: averting deprivation (e.g. savings clubs, social insurance)
- Promotive role: enhancing incomes and capabilities (e.g. inputs)

Over the past years, a new dimension has been added – transformative: which sees social protection as a mechanism to forge societal transformations such as social equity and inclusion, empowerment, and human rights.

The definition of what constitutes social protection has been an issue of much discussion. In a 'simple' definition, Harvey et al. (2007) state that social protection is concerned with protecting and helping those who are poor and vulnerable, such as children, women, older people, people living with disabilities, the displaced, the unemployed, and the sick. In a joint statement on universal social protection in June 2015, the ILO and the World Bank called on world leaders to promote universal social protection, which they define as 'the integrated set of policies designed to ensure income security and support to all people across the life cycle – paying particular attention to the poor and the vulnerable' (ILO and World Bank 2015, 1).

Freeland (2012) argues that there has been a proliferation of slightly ridiculous terms such as 'social springboards', 'social trampolines', 'safety ropes', and 'safety ladders' to try and capture metaphorically what social protection does. He asks people to differentiate social assistance from safety nets and from social protection. For him 'social assistance' defines a subset of social protection, comprising those social transfers that are non-contributory (i.e. which are funded from general government revenue, rather than from specific contributions by individuals). The key characteristics of social assistance are regularity, predictability over the long term, government ownership, and entitlement. 'Safety nets', on the other hand, is a term originally coined by the Bretton Woods institutions in the 1980s and 1990s that refers to temporary measures to catch those who were transiently made vulnerable through structural adjustment and liberalisation. Most people continue to consider safety nets in this original, narrower definition as a temporary social transfer project, usually operated for a finite period and often outside of government structures. A social protection system, however, is a much broader concept. Devereux and Sabates-Wheeler (2004, i) define social protection as 'all public and private initiatives that provide income or consumption transfers to the poor, protect the vulnerable against livelihood risks and enhance the social status and rights of the marginalised with the overall objective of reducing the economic and social vulnerability of poor, vulnerable and marginalised groups'. A similar definition of social protection is provided by Simson (2020) who describes it as a system of formal (and sometimes informal), public (and sometimes private) interventions that aim to reduce social and economic risks, vulnerabilities, exclusions, and deprivations for all people and facilitate equitable growth. As such, social protection may be best characterised as an umbrella which embraces both social assistance and safety nets.

This chapter embraces this holistic view of social protection but with some qualifications. First, social protection appears to be an exclusionist concept: the adjective 'social' tends to emphasise the social dimension of the interventions while, as seen from the definition above, social protection encompasses a wide range of interventions in the social (the minor part), human, physical, natural, financial, and food security dimensions. The noun 'protection' suggests a sense of weakness, powerlessness, and fragility, masking the capabilities that the group possesses, and excluding any consideration of and emphasis on human agency. As argued by Artur (2011), if the beneficiaries are portrayed as weak, passive, and powerless, then the interventions will tend to take a top-down approach with little understanding of local capabilities, knowledge, and motivations. It would be more productive to understand vulnerability as embedded in politics and power relations. As I will show from the Mabote case study, social protection interventions are indeed a battlefield; they take place in contexts where actors with different worldviews, interests, and power relations meet, dispute, and negotiate open-ended processes and outcomes.

Finally, the process of targeting the beneficiaries for social protection is also exclusionist. In an era where the world is calling for inclusion, equal rights, the cross-fertilisation of ideas, beliefs, norms, and worldviews, where kids with special needs deserve to be schooled along with other children, and where people regardless of their background, religion, race or nationality deserve the same rights and opportunities, we should stop (positively) segregating the so-called weak and envisage more inclusive empowerment approaches. As argued later, segregation has done more damage to local communities than good by excluding the vulnerable from internal and external networks which support them more than short-term, un-contextualised, and limited external interventions.

Social protection in Mozambique: so many to protect, so little to provide

Mozambique is considered the world's 10th most vulnerable country to disasters (INGC 2017). Located in Southeast Africa, along the Indian Ocean, the country is affected, on average, by one disaster of great magnitude every year (INGC et al. 2003) and ranks third on global weather-related damage, following Bangladesh and Ethiopia (Buys et al. 2007). Floods, epidemics (derived partly from flooding), cyclones, and drought comprise most major hazards. To deal with these disasters, the government in Mozambique has formulated a number of policies and strategies. More particularly, in 2012, it approved the National Strategy for Climate Change Adaptation and Mitigation which outlines strategic objectives and interventions for climate change adaptation and mitigation. In 2014, a national disaster law was passed which provides legal mechanisms for disaster prevention and response and, in 2017, the government approved a master plan for DRR aligned with Sendai Framework and the UN's SDGs.

Poor people are at risk of being the most affected by climate change and variability due to a combination of social and climatic factors that exacerbate their vulnerability. The occurrence of climate shocks and stressors, such as unseasonal droughts, cyclones, or floods negatively affects rural livelihoods, reducing, in turn, people's wellbeing. It is against this background that the government of Mozambique approved, in 2015, the national strategy for basic social protection

(*Estratégia Nacional de Protecção Social Básica* – ENSBB 2016–2024). The implementation of ENSSB is based on four key programmes:

1. PSSB (*Programa Subsídio Social Básico* – Basic Social Subsidy Programme): This is a monthly cash transfer programme targeting the so-called 'permanently' non- capable/handicapped people such as the elderly, people with physical handicaps, and people chronically ill. The amount received varies from 540 MT (about USD $9) to a maximum of 1000 MT (about USD $16) depending on how big the size of the beneficiary's household.[3] This is a permanent programme that assists people up to their deaths.
2. PASD (*Programa de Apoio Social Directo* – Direct Social Support Programme): This is an in-kind (mainly food provision) programme targeting people and households with specific needs at the time. For example, it targets people suffering from nutritional deficits or a household that has experienced temporary misfortune. The support, equivalent to a monthly payment of 2500MT (about USD $40), ceases once the beneficiary recovers from the problem he/she faces.
3. PASP (*Programa de Acção Social Produtiva* – Productive Social Support Programme): This programme targets the poor and vulnerable households who are fit enough to be involved in (a) public work and (b) livelihoods/asset development. It is a seasonal programme which lasts four months in rural areas and six months in urban ones. For public works, the recipients work four days a week for four hours for a monthly subsidy of 1050 MT (about USD $17). Unfortunately, due to limited funding, the livelihood component has never been implemented and the focus has only been on public work.
4. PSSAS (*Programa de Serviços Sociais de Acçao Social* – Social Services Programme): This programme provides assistance to orphanages, centres for the elderly, and schools (in order to keep poor and vulnerable kids in schools for instance) and health support (in order to allow the poor and vulnerable to continue receiving treatments).

By 2014, social protection programmes had directly assisted 427,000 people, an increase of 60 per cent compared to 2010 (254,000 beneficiaries). The overall budget allocated to social programmes increased from 0.22 per cent of the GDP in 2010 to 0.51 per cent in 2014. By 2024, the government expects to reach approximately 3,352,515 direct beneficiaries and a further 8,274,789 indirectly, comprising indirect assistance to about ¼ of the total population of Mozambique. To achieve this, the government expects to allocate about 2.23 per cent of the GDP, or about five times more than current expenditure (GoM 2015).

To be fully effective, the amounts expended, and the number of people included in the social protection system is still too limited. The highest amount an individual can receive is from the in-kind support of 2500MT (about USD $40) a month. The national worker's association (*Sindicato Nacional de Trabalhadores*) estimates that the monthly basic food basket costs about 14,000MT (circa USD $230), or six times more than the maximum currently provided (IESE 2018).

Moreover, it must be stressed that the total number of people living under the poverty line in Mozambique is about 13.3 million (46 per cent out of 29 million). The government also estimates that: 43 per cent of children suffer from acute malnutrition; 13 per cent of children are orphans; 14.3 per cent of girls below the age of 15 and 48.2 per cent below the age of 18 have entered early marriage; 46.5 per cent of people 15-19 years old and 29.6 per cent between the ages of 20 and 24 are unemployed (2014/2015); about 15 per cent of the population between the ages of 25 and 34 live with HIV; and only 12 per cent of the over 60s are in a formal social protection scheme (GoM 2015). The amount spent on social protection either using the period before 2014 (0.51 per cent of GDP) or even the expected 2.23 per cent in 2024 is just too little compared to other countries, with fewer people in need of government support. For example, France spends 31.2 per cent of GDP, Germany 25.1 per cent, and the UK 20.1 per cent. Even other African countries spend more: South Africa 4 per cent, and Malawi and Ethiopia 4.5 per cent.[4] This is a reason to say: so much to protect, so little to provide. In Mozambique, informal social protection mechanisms will continue to be the main safeguard of the majority of those in need of it. In this context, it makes no sense to ignore vulnerability and focus (solely) on resilience. And there are ample reasons and evidence to argue that vulnerability still matters!

Social protection in Mabote district: from poverty to vulnerability and back again to poverty

This section presents a case study based on Mabote district, Southern Mozambique. It highlights how poverty and vulnerability interact and continuously shape one another over time making it hard to sustain so-called resilient development. Data for the case study was collected in 2017 through both qualitative and quantitative methods. 330 households were randomly selected for survey and researchers carried out 11 semi-structured interviews with key informants and held five focus group discussions (FGDs), two of these were with women, two with men, and one mixed group. A literature review about the field site was conducted, as was participant observations for a period of over a month. The research was part of a newly funded project by Irish Aid intended to promote adaptive social protection (ASP), an approach that attempts to link social protection to climate change adaptation. The main objective of the research was to provide an academic background analysis in support of effective project implementation. The main research question concerned how interventions on social protection can improve, sustainably, the adaptive capacity of beneficiaries. The results were used to inform project (re)design and implementation.

Brief District Profile: gendered poverty and vulnerability

Mabote district (Figure 9.1) is located in the northern part of Inhambane province, about 363 km from the provincial capital of Inhambane. Its surface area

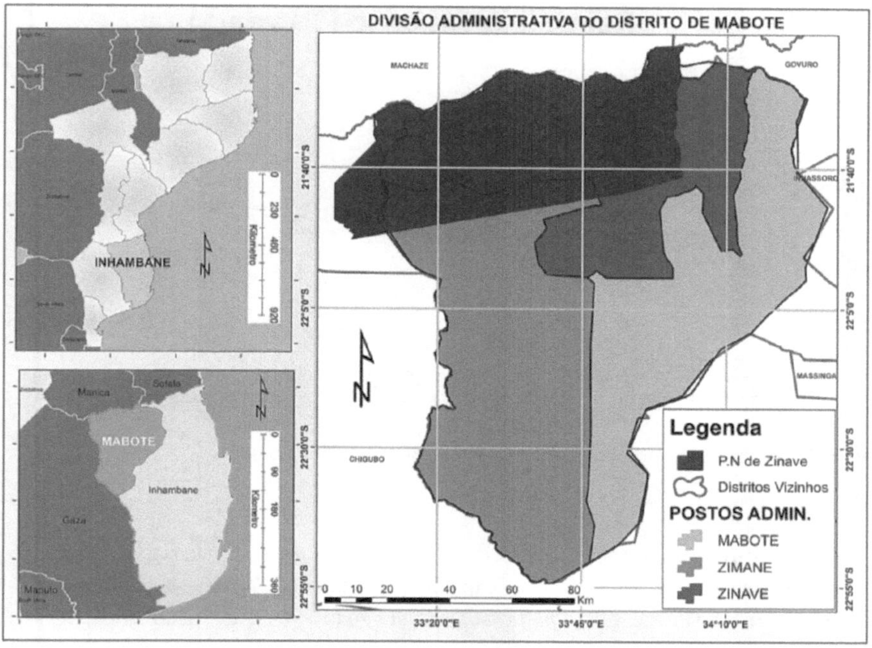

FIGURE 9.1 Geographical location of Mabote district and its administrative division.

of 14,577 km^2 is divided into three administrative posts, *postos administrativos* (Mabote-Sede, Zinave, and Zimane), and eight localities, *localidades* (Chitanga, Mabote-Sede, Papatane, Mussengue, Maculuve, Tanguane, Zinave, and Benzane). According to the National Institute of Statistics, Mabote had 52,719 inhabitants in 2017. The distribution of this population is scattered, with most people, 24,673 (46.7 per cent), living in the vicinity of the district town, Mabote Sede. The remaining population live either in Zinave, 19,875 (37.7 per cent) or Zimane with just 8,171 (15.6 per cent).

Mabote has a dry subtropical climate with two seasons – dry and rainy. The dry season is dominant over a period of eight months from March to October. Annual precipitation is very low (about 600 mm) and about 90 per cent of it occurs in just four months (November-February). In 2015 and 2016, rainfall was exceptionally low, reaching only 284.3 mm and 430.13 mm respectively – well below average. The resultant drought led to drastic agricultural losses and a need for humanitarian aid (SDAE 2017). Overall, as shown in Figure 9.2, rainfall in Mabote shows considerable inter-annual variation with totals ranging from a high of 1297.5 mm to a low 251.2 mm. Data for the period 1988–2016, however, shows a declining trend in annual rainfall with, over the past 10 years, only four years when precipitation rates were equal to or above the annual average, and with the remaining six years having rainfall below that figure.

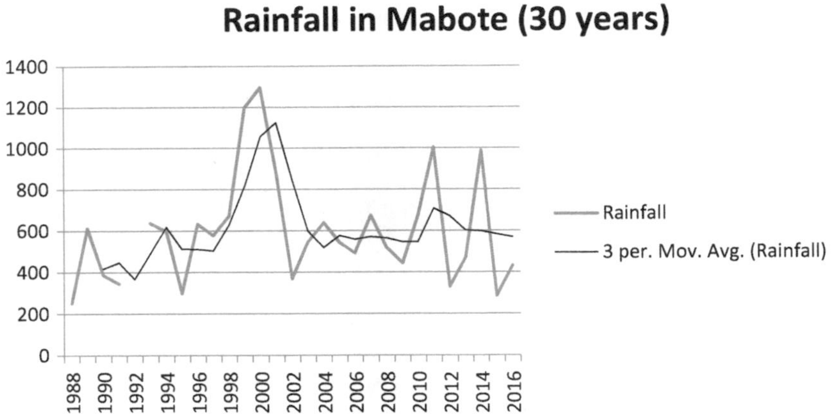

FIGURE 9.2 Annual rainfall in Mabote 1988–2016. (SDAE 2016)

Vulnerability mapping with FGDs (through scorecards that examined the exposure, sensitivity, and adaptive capacity) shows that Zimane is the most vulnerable administrative post (*posto administrativo*) followed by Mabote-Sede and then, least vulnerable, is Zinave. In terms of the localities (*localidades*), Table 9.1 provides an overview of the vulnerability intensity across the district. Overall, three localities are ranked in the scorecard exercise as high, two as moderate, and three as low.

Rainfall agriculture, livestock rearing, and extraction of forest products constitute the main livelihood sources in Mabote. According to SDAE (2017), drought-resistant crops such as millet, sorghum, cassava, and beans are the most preferred. In the 2016/2017 agricultural season, millet contributed 27 per cent of the overall production, peanuts 22 per cent, sorghum 21.4 per cent, cassava 14

TABLE 9.1 Vulnerability mapping by score cards (exposure, sensitivity, and adaptive capacity)

Vulnerability intensity by comparing administrative posts (postos administrativos)		*Vulnerability intensity by comparing localities (localidades)*	
Administrative Post	Vulnerability intensity	Locality	Vulnerability intensity
Mabote sede	Moderate	Sede	Low
		Chitanga	Moderate
		Papatane	High
Zinave	Low	Maculuve	Moderate
		Mussengue	Low
		Tanguane	High
Zimane	High	Zimane	low
		Benzane	High

per cent, beans 10 per cent, and maize just six per cent. Cashew nut is the main cash crop and the price that season was exceptionally good, reaching nearly USD $1 per kg. Livestock is dominated by cows (23,128 animals or approximately 45 per cent of all livestock in the district – an average of two cows per household) and chickens (15,435 birds or about 30 per cent of livestock in the district – an average of two per three households). Other animals reared include goats (8,633 animals – an average of one animal per two households) and pigs (1,176 animals – an average of one animal per 10 households). Migration (to South Africa) and casual labour such as selling (mainly in the cities of Vilanculos and Inhambane) are other major livelihood sources.

Poverty rates are exceptionally high in Mabote. As part of the vulnerability mapping, we carried out a survey (Sitole et al. 2017). This found that 62 per cent of households were below the poverty line. In contrast, the official poverty line in 2015 was estimated at 48.6 per cent in Inhambane province and nationally at 46.1 per cent. The multidimensional poverty line used in our survey scored 50 per cent, well above the provincial figure at 43 per cent but below the national figure at 55 per cent. Only approximately five per cent of residents had cement houses, seven per cent had electricity, only about 12 per cent had access to extension services, and 15 per cent to (informal) financial services. 58 per cent of respondents reported lost crops due to drought and 55 per cent mentioned losses due to a combination of droughts and flash floods. Our survey also found that by 2016 about 53 per cent of people in Mabote needed aid to survive.

Disaggregated data further show that poverty is gendered. More women live in unsafe houses compared to men (76 per cent against 64 per cent).[5] Illiteracy rates are higher among women than men (52.7 per cent against 21.3 per cent). Men tend to have better agricultural plots than women (14.1 per cent against 11 per cent) that, on average, are also bigger than those cultivated by women (2.9 ha against 1.7 ha). Male-headed households possess more livestock than female-headed households – in terms of cattle 43.5 per cent against 10.2 per cent, goats 24.7 per cent against 7.4 per cent, and chickens 48 per cent against 31.8 per cent. On average, too, men own more animals than women (2.1 Total Livestock Units against 0.7) and women's assets are worth less than men's assets (35,170.68MT against 56,040.13MT). Overall, it appears that women are poorer than men (79 per cent of female-headed households were below the consumption poverty line against 42 per cent of male-headed households), and 65 per cent of female-headed households are under the multidimensional poverty line that assesses multiple deprivations compared to 33 per cent of male-headed households.

A key factor affecting women's wellbeing is related to migration. In drylands such as Mabote, male migration becomes a norm to the extent that not migrating is equated with not being 'macho enough'.[6] Every adult male has migrated either to South Africa or to somewhere else outside his community at least once in his lifetime. This happens when men are still young and energetic and allows them to accumulate the assets necessary to contract marriage. It is also a form of social protection working, and as such, is a key strategy for resilience. In a separate

TABLE 9.2 Number of years men had spent in South Africa (SA) and their employment

Respondent	*Years in South Africa*	*Job*
1	10 (5 trips of 2 years each)	Mining
2	12 (annual trips, mostly back for Christmas and New Year)	Bricklayer
3	12 (6 trips of 2 years each)	Mining
4	12 (6 trips of 2 years each	Mining
5	14 (7 trips of 2 years each	Mining
6	12 (annual trips there and back)	Electrician
7	8 (4 trips of 2 years each)	Mining
8	10 (2 trips of 2 years each)	Private security
9	10 (5 trips of 2 years each)	Railway maintenance
10	12 (4 trips of 3 years each	Chef and mining
11	4 full years	Mining
12	10 (5 trips of 2 years each)	Mining
13	18 (9 trips of 2 years each)	Mining
14	14 (7 trips of 2 years each)	Mining
15	10 (5 trips of 2 years each)	Mining
16	8 (4 trips of 2 years each)	Carpenter

male-only FGD held in Benzane attended by 16 men, we asked respondents whether they had migrated somewhere in their lifetimes. On average, the men had spent 11 years outside their communities (see Table 9.2).

Long periods of male migration place a huge burden on the women left behind. They are tasked with productive, reproductive, and community roles which otherwise would be shared or assumed by men. Moreover, young migrant men engaged in non-agricultural and pastoral employment, such as mining, adds little to the skills and knowledge they will need later. Over time, too, they lose what farming skills they had, and their knowledge becomes outdated, rendering them literally useless for performing the key livelihood support activities required on return to their communities. Many male migrants also fail to send home their remittances, thereby increasing the burden on women who now have more people to take care of and support. In a women-only group discussion, they also complained that by 'doing nothing' men become psychological unstable, overstressed, and heavy drinkers.[7] Gender-based violence is frequent. Men's lack of key agricultural skills was illustrated in the FGD held in Benzane. We asked respondents what kind of interventions would help them to better deal with recurring droughts and flash floods. Two response streamlines emerged. Women asked for irrigation schemes, new seed varieties, livelihood diversification, and better integration into the market through financial support and financial literacy. Men, on the other hand, wanted jobs: 'we need jobs; there are no jobs here'!

Assets accumulated during migration such as acquiring livestock become a male (social) bargaining tool and may partly explain why cattle numbers are increasing in Mozambique's drylands. As observed by researchers (i.e. Newitt 1995; Hedges 1977), cattle are very relevant for marriage and other social networks in South Mozambique. Newitt (1995) even suggests that former male migrants may no longer be able to support themselves if left alone at an older age because they have lost the necessary local living skills required. Therefore, in addition to wealth accumulation when younger, multiple marriages offer men a safety net to survival when older. Overall, women who are already burdened with multiple tasks and limited support, are further stretched by climate change-related impacts that tend to increase their vulnerability and poverty. Therefore, there is a need to disaggregate climate change impacts and to always pursue gender and climate-sensitive interventions.

Social protection in Mabote: excluding the unprotected and protecting the protected

By 2017, the government of Mozambique was assisting people in Mabote through its social protection system with 2,869 beneficiaries and another 5,000 on the waiting list.[8] The government has designed a nine-step targeting system to select beneficiaries which involves[9]:

1. Approving an annual budget at the central (national) level
2. Informing each province of the funds it has been allocated and so the number of beneficiaries that can be enrolled for social protection
3. Each province informing the district of the number of beneficiaries it can enrol
4. Each district distributing the number of beneficiaries across *postos administrativos* and *localidades*
5. Local community leaders composing a list of poor and vulnerable people that is supposedly discussed and agreed in a community meeting before being sent to the district level
6. Sending this list to a central level database for assessment
7. Dispatching, on an annual base, teams from the provincial or regional level to assess 'in loco' the plight of nominated beneficiaries and to collect relevant socio-economic data that are used to calculate the PMT (Proxi Mean Test)
8. Calculating a PMT and sending the results back to each district so that the list of beneficiaries can be reshaped accordingly
9. Placing beneficiaries in different programmes based on their characteristics (i.e. fitness for work, elderly, etc.)

Out of the 2,869 beneficiaries in Mabote, 1,857 (65 per cent) were enrolled in the Basic Social Subsidy Programme. The PSSB is a cash transfer to elderly, handicapped, and chronically ill people. By 2017, the monthly transfer amounted to 310MT (circa USD $5) for a one-person household, 390MT (circa USD $6.5) for a two-persons household, 460MT (circa USD $7.6) for a three-person household, 530MT (circa USD $8.9) for a four-people household, and 620MT (circa USD $10) for a five-person or more household.

The PASD or Direct Social Support Programme that provides in-kind support assisted 150 people (5 per cent) and provided each person with: 1l of oil,

1kg of salt, 2kg of peanuts, 4kg of maize flour, 5kg of rice, and 5kg of beans as well as a soap and matches. The PASP or Public Works Programme involved 862 people (30 per cent) who received 650MT/month for working four hours/day for four days over four months (4/4/4). The PSSAS or Social Services Programme provided food to feed 11,797 school students, 338 teachers, and 74 community members who acted as cooks.

There have been complaints about how the targeting system works in practice. Locals, for instance, claim that there are few public hearings over who is selected as beneficiaries and accusations of favouritism with local leaders choosing people from their own closest networks. Others maintain that the PMT has a large in-built design error, as there is no perfect correlation between the observed proxies and real household consumption. As a result, even prior to households being surveyed, a high proportion of potential beneficiaries may be excluded (Transform 2017, 15).

In this chapter, I do not intend to discuss how the targeting system works to select beneficiaries *per se* but rather how the process in Mabote tends to exclude those most in need and protect those who are already, so to speak, protected. Table 9.3 provides the geographical location of the 2,869 beneficiaries directly involved in the formal social protection system across all districts in Mabote as of 2017 and compares that with the vulnerability index discussed previously.

The table shows two very important elements: the first is that Mabote-Sede, that scored low on the vulnerability index, hosts nearly half of all targeted beneficiaries, while Zimane, that scored high on the vulnerability index, includes less

TABLE 9.3 Social protection and vulnerability

Administrative post	*Locality*	*Inhabitants*	*PSSB*	*PASD*	*PASP*	*Total*	*Vulnerablity index*
MABOTE SEDE	SEDE	24673	532	115	255	902	Low
	CHITANGA		236	5	65	306	Moderate
	PAPATANE		95	7	83	185	High
TOTAL		46.70%	**863**	**127**	**403**	**1393**	48.5% of the targetted
ZINAVE	MACULUVE	19875	312	6	110	428	Moderate
	MUSSENGUE		84	5	59	148	Low
	TANGUANE		146	4	90	240	High
TOTAL		37.70%	**542**	**15**	**259**	**816**	28.4% of the targetted
ZIMANE	ZIMANE	8171	191	5	200	396	Low
	BENZANE		261	3	0	264	High
TOTAL		15.60%	**452**	**8**	**200**	**660**	23% of the teragetted
GRAND TOTAL		52719	**1857**	**150**	**862**	**2869**	
						Total low	1446 (50.4%)
						Total moderate	734 (25.6%
						Total high	689 (24%)

than a quarter of those targeted. Second, the total number of people selected from all regions considered less vulnerable encompasses more than 50 per cent of the beneficiaries, while those living in areas considered highly vulnerable total just 24 per cent. The table shows that there is clearly a mismatch between vulnerability and social protection interventions. It appears that the system is protecting those who are not, within the limits of the available resources, most in need of protection.

Besides comparing the geographical location of the beneficiaries, I also analyses the technical support provided by the government to increase resilience by examining where the extension services are more concentrated. Table 9.4 provides a synthesis of the analysis.

Like the social protection targeting system, it appears that technical support also tends to be directed toward less vulnerable areas rather than highly vulnerable ones. It is claimed that operational costs are too high in the most vulnerable areas and that they are far from the provincial centre and access to them is difficult. Rather than expending funds on the most marginal areas, government agencies prefer to focus on those areas where access is easiest, that are geographically closest, have more assets – and are less vulnerable. In short, it might be argued that the government cares less for the poor and most vulnerable.

Lastly, it is important to look at the type of support provided by the extension services. The analysis compared the seeds provided by government programmes in the recovery process to the types of seed local people wanted. Figure 9.3 presents a plate of seeds brought by local people that contains the seeds of interest to them. In this picture, there are five types of seed: millet, sorghum, cowpea, cucumber, and watermelon. As shown in Table 9.4, the seeds provided in the drought recovery programme had little relation to locals needs. Instead, the seeds

TABLE 9.4 Extension services and vulnerability

Administrative post	*Locality*	*# technical staff with secondary or higher school training*	*# technical staff with basic/lower school training*	*Total # of technical staff*	*Vulnerability index*
	Mabote-sede	1	2	3	low
Mabote sede	Chitanga	1	1	2	Moderate
	Papatane	0	2	2	high
	Mussengue	2	1	3	low
Zinave	Maculuve	0	1	1	moderate
	Tanguane	0	2	2	high
Zimane	Zimane	3	0	3	low
	Benzane	1	1	2	high
Total staff		8	10	18 +director	
Total staff in low vulnerability areas					9
Total staff in highly vulnerability areas					6
Total highly qualified staff in low vulnerability areas					6
Total highly qualified staff in highly vulnerability areas					1

FIGURE 9.3 A plate of locally desired seeds containing millet, sorghum, cowpea, cucumber, and watermelon.

TABLE 9.5 Seeds and their relevance for local farmers

Crop	*Amount of seeds donated (Tons)*			*Local ranking of the crop*
	By the government	*By partners*	*Total*	
Millet	0	0	0	1st
Peanut	0	9.8	9.8	2nd
sorgum	0.45	0	0.45	3rd
Cassava	0	0	0	4th
Common Beans	0	10.7	10.7	5th
Maize	1.7	19.02	20.7	6th
Sweet potato	6	0	6	7th
Vegetables	0.003	0.104	0.1	8th

provided focused on maize, common beans, and vegetables that are less relevant for drought-prone areas such as Zimane – although they are more commonly used in the less drought-prone areas around Mabote-Sede. Again, it appears that the interventions benefited the less poor (Table 9.5).

Bringing back vulnerability under adaptive social protection

Over the past few years, a growing number of scholars have argued that there is much to be gained from linking social protection to climate change adaptation and disaster risk reduction in what has been coined by Davies et al. (2008) as Adaptive Social Protection (ASP). ASP allows those individuals, households, and communities that are vulnerable and part of social protection schemes to become more 'resilient' to shocks, to reduce poverty, and to promote human development (Béné et al. 2013; Kuriakose et al. 2012; Devereux and Sabates-Wheeler 2004). As a result, ASP projects and programmes have been established in different parts of the world. For example, the World Bank's Sahel Adaptive Social Protection Programme includes six Sahel countries as of 2014: Burkina Faso, Chad, Mali, Mauritania, Niger, and Senegal. It has become a flagship programme to showcase ASP and garner buy-in from different donors and countries.

The approach in Mabote: focusing on transformative adaptation

ASP, however, is still an approach in the making. At its core is the link between social protection and climate change adaptation. Those targeted for social protection should be the first to receive climate change adaptation interventions. Likewise, social protection interventions should be climate sensitive and increase the adaptive capacity of targeted groups. This approach addresses, comprehensively, the most vulnerable and is one where the concept of vulnerability still clearly matters.

In Mabote, Irish Aid is supporting an initiative led by local government and that involves the International Institute for Environment and Development and Eduardo Mondlane University. The initiative also goes beyond just linking climate change adaptation to social protection and addresses a complex array of underlying factors that create vulnerability in what I call Transformative Adaptation (TA). The initiative is called PRIORIZE and focuses on the following components:

1. Strengthen local government's capacity to plan and implement Adaptive Social Protection: the design and implementation of a Local Plan of Adaptation (LPA) to climate change that embeds adaptive social protection; approved interventions based on the capacities and resources available to the most vulnerable; and mainstreaming the resultant planning model by the district government.
2. Support the livelihoods of the most vulnerable and link local producers to markets: the implementation of smart agriculture with a focus on locally relevant

crops and technologies, local chicken value chains, cashew nut value chains, and honey value chains; water provision; gender and social transformation; and management of natural resources with a focus on forestry and biodiversity – all of which must be carefully discussed and approved by potential ASP beneficiaries.

3. Strengthen local civil society capacity to support government and local communities: provide technical, financial, and organisational support to a network of local organisations under the umbrella of ADELMA (*Associação para o Desenvolvimento Local de Mabote* – Association for the Development of Mabote) to make it a future development hub for both the government and local community.
4. Build platforms for continuous learning and improvement: establish national, provincial, and district multi-stakeholder platforms (all institutions that deals with climate change, disaster risk reduction, and social protection at their respective levels) that meet periodically to access the intervention and jointly propose any changes required so ensure that the most vulnerable remain the principal focus of interventions.
5. Tackle gender and social inequalities: address gender disparities as an integral part of all activities by assessing all interventions against their benefits or impacts on women and men.
6. Disseminate and advocate in support of ASP: utilise a webpage, WhatsApp, Facebook, and organise seminars.

Lessons so far in vulnerability reduction

PRIORIZE is a one-year pilot project that started with scoping studies in mid-2017 before introducing the practical interventions outlined between mid-2018 and the end of 2019. So far, the interventions have led to a shift in orthodox thinking.

1. Focus on assets development: social protection schemes use a cash transfer approach. Indeed, the tiny amounts of money that beneficiaries receive are mostly spent (if wisely and more often by women) on basic needs such as food and clothing and hardly ever on long-term investments. Asset development offers the prospect that interventions might sow the seeds for a longer lasting vulnerability reduction. The assets provided need to match people's abilities and capacities, and the environment where they live. PRIORIZE focuses: on poultry (chicken) as this is the asset most commonly owned and reared by the most vulnerable, and is also more resilient to drought if compared to cattle; on cashew nut production as it is a drought-tolerant cash crop with which the participants are familiar; and on beekeeping and forest protection to help fight climate change and because the more vulnerable are dependent on forest products for a livelihood, especially during drought.

2. Shift from individual to collective action: beneficiaries must be active members of their communities and be fully involved in interventions. Current social protection practices that only provide individual cash or goods transfer do not develop the collective action required to address issues such as climate change. In Mabote, asset development is through groups and cooperatives engaging in cross-learning and collective action. It is important, therefore, to empower local organisations and to support the capacity-building potential of ADELMA.
3. Integrate the marginalised poor into the market: as current approaches do not focus on asset development, there has been little integration of beneficiaries into the market. Beneficiaries contribute only marginally to the market as current programmes aid recipients on an individual basis and do not attempt to integrate producers into a wider market system. The current initiative is working to enhance participation in the market through integrating beneficiaries into value chains and supporting the training necessary to achieve this such as financial literacy, post-harvest and processing practices, and value addition.
4. Create multi-stakeholder platforms: under a transformative approach like that pursued by PRIORIZE, there is no need to pre-define entry-points (i.e. the Ministries of Social Development, Climate Change, or Environment). A preliminary assessment would determine the key challenges, who are the key stakeholders, and under which leadership the most effective transformations could be achieved. A pre-definition of a leading institution has brought frustration to many either because the leading institutions have failed to produce meaningful transformation at the local level or to bring together a stakeholder platform that can ensure the transformations needed. The leading institution should emerge from a consensus among the key stakeholders involved and this role should be rotated over time.
5. Ensure open-ended, flexible, and locally driven agendas: currently ASP practices involve pre-defined sets of interventions in agriculture (such as 'climate smart' agriculture practices) and other livelihoods. Technical staff are much happier disseminating these technologies. But they tend to be inflexible in their approach and often have a poor understanding of local dynamics and power relations. Flexibility and a good understanding of what works locally is key to addressing the root causes of what systematically makes people vulnerable and to forging more long-lasting impacts. Local Adaptation Plans have a vital role to play in this.
6. Facilitate direct implementation rather than multiple intermediaries (gatekeepers): most ASP interventions fail to achieve their goals because too little money reaches beneficiaries. A conglomerate of gatekeepers that includes international and local NGOs, government bodies, experts, and service providers take too big a 'share of the cake' (GHA 2018). Dispensing with these gatekeepers would provide much-needed resources for the transformations required to lift people and communities out of vulnerability. PRIORIZE has been able to direct funds directly from the UK to districts and communities in Mabote and to the interventions they are supposed to fund.

Key lessons learnt in implementing transformative adaptation

There have been multiple challenges in attempting such a radical shift through implementing TA. The most important ones are:

1. Timeframe: as TA works within a context where norms, beliefs, and practices are entrenched at international, national, and local levels, changes tend to be incremental and require longer time commitments.
2. Resources: the platform building for successful transformation as envisaged here requires good facilitation skills and staff that understand the philosophical underlying principles. Unfortunately, staff with that sort of analytical capacity and commitment are hard to find. In addition, under TA results are not easily tangible and interventions are implemented in areas and with people who are hard to reach and to speak with. For those more interested in ticking the box and looking for a quick fix, this approach might not be appealing.
3. Power politics: TA is about shifting who holds the power through more flexible and bottom-up approaches. Such a shift provokes conflict as it threatens the status quo and deprives elites from capturing the resources they need to maintain themselves in power. To counter this, well-functioning platforms with leaders versed in conflict management skills and group dynamic understanding are crucial component of the programme from its start.

Concluding remarks

This chapter began with the claim that climate change adaptation, disaster risk reduction and social protection interventions fail to lift people out of poverty and vulnerability because they use inappropriate targeting tools and interventions. The global political, economic, and social context over the past few decades has, to some extent, increased people's vulnerability rather than enhancing their resilience. Global developments have been pro-rich, inequalities have increased all over the world, and countries like Mozambique are left struggling to come up with alternative approaches that address the resultant multiple crises and help their most vulnerable citizens deal with the impacts of climate change. This case study from Mabote district represents in microcosm a trend that assails many people in many states worldwide: gendered poverty, interventions that benefit the already better off in communities, and programmes that do not address the root causes of continuous poverty and vulnerability. In such a context, the capacity of poor people is only further eroded, leaving them with limited room to manoeuvre when it comes to climate change impacts and other stressors. It matters, therefore, to address their vulnerability by looking at alternative pathways that are outside of hegemonic thinking and current practices. As I have shown, there are alternative options available and already being tested. PRIORIZE is one such initiative that attempts to reverse the current trend by proposing innovative shifts in thinking, planning, implementation, and evaluation. It focuses on the most vulnerable – their capacity and resources – and creates around them a supportive environment that promotes inner power and creativity. It is a 'think big but start small' approach that offers prospects for alternatives in a world that praises uniformity and homogeneity more than *la art de la localite* – to paraphrase the Dutch scientist Van der Ploeg (1989) – that is required for adaptation and social protection.

Notes

1 For further details on the Mozambican financial scandal, please refer to Banco Mundial 2016, Francisco and Semedo 2016 and 2017, and Aiuba 2019.
2 According to the Bank of Mozambique, the economy decreased by 3.25 per cent in the second trimester and by 1.09 per cent in the third trimester of 2020 as a result of the COVID-19 pandemic (Banco de Moçambique 2020). http://www.bancomoc.mz/ accessed November 28, 2020.
3 This is an updated amount that started in 2018. Before this the amounts were much lower and will be presented later.
4 https://stats.oecd.org/Index.aspx?DataSetCode=SOCX_AGG
5 Mainly made from local material such as grass, trees, soil and robs.
6 Excerpts from FGD in Benzane, April 2017.
7 Excerpts from FGD with females in Chitanga, April 2017.
8 Interviews at Mabote delegation of National Institute of Social Protection (INAS) April, 2017.
9 Interviews at National INAS, March 2019.

References

Adger, Neil. 2000 *Social & Ecological Resilience: Are They Related?* Progress in Human Geography 24 (3): 347–364.

African Development Bank. 2019. *African Economic Outlook 2019.* African Development Bank Publications. Avenue Joseph Anona, 01 BP 1387 Abidjan 01.

Aiuba, Rabia. 2019. *A Economia de Moçambique está a recupera?* Destaque Rural 58. Maputo, Moçambique.

Allister McGregor, J.. 2011. *Reimagining Development through the Crisis Watch Initiative.* IDS Bulletin 42(5), 17–23.

Arndt, Channing and Kristi Mahrt. 2017. *Is Inequality Underestimated in Mozambique? Accounting for Underreported Consumption.* WIDER Working Chapter 2017/153.

Artur, Luís. 2011. *Everyday Practices of Disaster response and Climate change adaptation in Mozambique.* PhD Thesis. Wageningen University and Research Centre. The Netherlands.

Banco de Moçambique. 2020. *Estatísticas.* Available at http://www.bancomoc.mz/ accessed November 28, 2020.

Banco, Mundial. 2016. *Actualidade Económica de Moçambique. Enfrentando Escolhas Difíceis.* The World Bank Group, Washington, DC.

Béné, Chris, Terry Cannon, Mark Davies, Andrew Newsham, and Thomas Tanner. 2013. *Social protection and climate change.* OECD-DAC Task Team on Social Protection, 15th ENVIRONET Meeting, Paris, 25–6 June.

Berg, Andrew and Jonathan D. Ostry. 2011. *Equality and Efficiency: Is There a Trade-Off Between the Two or Do They Go Hand in Hand?* Finance & Development, September: 12–15.

Burns, Matthew. 2011. *School Psychology Research: Combining Ecological Theory and Preventive Science.* School Psychology Review 40, 132–139.

Buys, Piet, Uwe Deichmann, Craig Meisner, Thao Ton, and David Wheeler. 2007. *Country Stakes in Climate Change Negotiations: Two Dimensions of Vulnerability.* The World Bank, Washington, DC.

Club of Mozambique. 2016. *Business news.* Available at https://clubofmozambique.com/news accessed November 20, 2016.

Davies, Mark, Bruce Guenther, Jennifer Leavy, Tom Mitchell, and Thomas Tanner. 2008. *'Adaptive Social Protection': Synergies for Poverty Reduction.* IDS Bulletin 39 (4), September. © Institute of Development Studies.

Devereux, Stephen and Rachel Sabates-Wheeler. 2004. *Transformative Social Protection.* Working Chapter 232. IDS, Brighton.

Feijo, João. 2018. *Pobreza, Diferenciação Social e (Des)alianças Políticas no Meio Rural.* Observador Rural 61, Maputo.

Francisco, António and Ivan Semedo. 2016. *De novo a questão de saldos rolantes na conta geral do estado. IDeIAS* 91, IESE, Maputo.

Francisco, António and Ivan Semedo. 2017. *A face do orçamento do estado moçambicano: Saldos de caixas são fictícios? IDeIAS* 93, IESE, Maputo.

Freeland, Nicholas. 2012. *Safety Net ∤Social Assistance.* Pathways´ Perspective on Social Policy in International Development. Issue No 1. June.

Gabinete de Reconstrução Pós Ciclone Idai (GREPOC). 2020. *Relatório de Reconstrução e Recuperação Pós Ciclone Idai e Kenneth.* Outubro 2020. Maputo, Moçambique.

GHA. 2018. *Global Humanitarian Assistance Report 2018.* Geneva.

GoM. 2015. *Estratégia Nacional de Segurança Social Básica 2016–2024.* Maputo, Moçambique.

Gradin, Carlos and Finn Tarp. 2017. *Investigating Growing Inequalities in Mozambique.* WIDER Working Chapter 2017/208.

Hadded, Lawrence, Naomi Hossain, J. Alliston Mc Gregor and Lyla Metha. 2011. *Introduction. Time to Reimagine Development?* IDS Bulletin 42(5), 1–12.

Harvey, Paul, Rebbeca Holmes, Rachel Slater, and Eduard Martin. 2007. *Social Protection in Fragile States.* ODI, London.

Hedges, David. 1977. *Trade and Politics in Southern Mozambique.* PhD thesis, University of London.

IESE. 2018. *Salário Mínimo e Custo de Vida em Moçambique.* IDEIAS, 5 de Setembro. Maputo, Moçambique.

ILO. 2017. *World Social Protection Report 2017–19 Universal Social Protection to Achieve the Sustainable Development Goals.* ILO Publishing. International Labour Office, CH-1211, Geneva 22, Geneva, Switzerland.

ILO and World Bank. 2015. *A Shared Mission for Universal Social Protection. Concept Note.* The World Bank Group, 1818 H Street NW, Washington DC, 20433, USA

INGC. 2017. *Plano Director para Redução do Risco de Desastres 2017–2030.* Maputo, Moçambique.

INGC, UEM and Fewsnet. 2003. *Atlas for Disaster Preparedness and Response in the Limpopo Basin.* Maputo, Moçambique.

Irish, Aid. 2017. *Irish Aid Social Protection Strategy.* Department of Foreign Affairs, Government of Ireland, Edinburg.

Krugman, Paul. 2009. *How Did Economists Get It So Wrong?* The New York Times Magazine. September 2, 2009.

Kuriakose, Anne T, Rasmus Heltberg, William Wiseman, Cecilia Costella, Rachel Cipryk, and Sabine Cornelius. 2012. *Climate-Responsive Social Protection*. Social Protection & Labor Discussion Chapter 1210.

McCulloch, Neil and Amit Grover. 2010. *Estimating the National Impact of the Financial Crisis in Indonesia by Combining a Rapid Qualitative Study with National Representative Survey*. IDS Working Paper 2010 (346), 1–39.

Newitt, Malyn. 1995. *A History of Mozambique*. Hurst & Company, London.

Nova, Yara. 2019. *Evolução dos Preços dos Bens Alimentares 2018*. Observador Rural 71, Maputo.

Ravallion, Martin. 2005. *Inequality is Bad for the Poor*. Policy Research Working Chapter. World Bank Group, Washington, DC.

Reliefweb. 2019. *OCHA Situation Report*. Available at https://reliefweb.int/country/moz accessed March 12, 2019.

SDAE. 2016. *Balanço Anual*. Mabote.

SDAE. 2017. *Balanço Anual*. Governo do Distrito de Mabote, Mabote, Moçambique.

Simson, Michel. 2020. *Introduction to Social Protection as Both a Developmental and Covid-19 Response*. EPRI, South Africa.

Sitole, Rogério, Luís Artur and Simon Anderson. 2017. *Estudo de Base para a Iniciativa Priorize-Ligação entre Protecção Social e Resiliência Climática*. Relatório de pesquisa, Universidade Eduardo Mondlane, Maputo, Moçambique.

The World Economic Forum–WEF. 2017. *Global Risk Report 2017*. Geneva.

The World Economic Forum-WEF. 2019. *Global Risk Report 2019*. Geneva.

Transform. 2017. "*Selection and Identification in Social Protection Programmes – Manual for a Leadership and Transformation Curriculum On Building and Managing Social Protection Floors in Africa*", available at http://socialprotection.org/institutions/transform.

Van der Ploeg, Jan douwe. 1989. *Knowledge Systems, Metaphors and Interface: The Case of Potatoes in the Peruvian Highlands* in: Long (ed.) *Encounters at the Interface. A Perspective on Social Discontinuities in Rural Development*. Wageningen Studies in Sociology, WSS 27, pp 145–164.

WIR. 2018. *World Inequality Report. 2018*. World Inequality Lab. Berlim.

Wisner, Ben, Pier Blaikie, Terry Cannon, and Ian Davies. 2004. *At Risk: Natural Hazards, People vulnerability and Disaster*. Routledge, New York.

World Bank. 2003. *World Social Protection Report: The World Bank's Approach to Social Protection in a Globalizing World*. Washington, DC.

World Bank. 2011. *The World Bank 2012–2022 Social Protection and Labour Strategy*. Washington, DC.

World Bank. 2017. *Umbreakable -Building the Resilience of the Poor in the Face of Natural Disasters*. Washington, DC.

World Bank. 2018. *Mozambique Economic Update – Shifting to More Inclusive Growth*. Washington, DC.

Yalcin, Erdal, Luisa Kinzius, and Gabriel Felbermayr. 2017. *Hidden Protectionism. Non-Tariff Barriers and Implications for International Trade. GED study*. Berlim, German.

PART III
Disaster risk creation

10

POWER WRIT SMALL AND LARGE

How disaster cannot be understood without reference to pushing, pulling, coercing, and seducing

Ben Wisner

Introduction

Current understanding of risk is limited by a failure to appreciate the role of power. Power relations have to be considered when attempting to understand risk perception and behaviour, local group risk reduction initiatives and participation in municipal safety programs, and also when addressing national and international risk management policy and risk creation. To illuminate risk-scape and process fully at the granular, fine scale of the individual and group as well as larger scales, power needs to be treated as central in a number of different, non-mutually exclusive, forms. Considering a very general definition of power as the *ability to influence someone else's behaviour*, or, to follow sociologist Talcott Parsons (1963), to *obtain one's intentions despite opposition*, disaster studies needs to enquire into sources and the exercise of power in the social, economic, political, administrative, and ideological domains.

I develop this argument in three parts. First, I discuss the lack of value attributed to life worlds and, at times, their invisibility. Here I must establish that experience and culture have a practical value in risk communication and also value for demystifying power relations that work at this scale, as well as enabling group solidarity and resistance against state ideological hegemony. Such push back against unwholesome and hurtful exercise of power demands understanding local drivers for change, local differentiation, and everyday power differences within communities. I'll also argue that there is value for people for whom the life world is the ground of their identity. Secondly, I explore ways that the life world is acknowledged but dismissed by top-down imposition of Western science or manipulated in order to gain acceptance of preconceived 'solutions' and 'projects'. Finally, I turn to the outright attack on life worlds by a variety of

DOI: 10.4324/9781003219453-13

'modernising' or 'rationalising' public (Scott, 1998), public–private, and private investments, especially those involving eviction and displacement of people.

Those familiar with my earlier work will realise that the 'pressure and release' ('progression of vulnerability' or 'PAR' framework) does not accommodate the life world or other ways of understanding the subjective and experiential elements of vulnerability and capacity in relation to natural hazard risks (Blaikie et al. 1994; Wisner et al. 2004, 2012). The PAR framework assumes that power is a function of access to capital, information, and other means of production (and means of destruction) as well as access to the state and its apparatus. While this is without doubt true, PAR does not make sufficient room for the manipulation of meaning and language by politicians and the state, by media, and by religious personalities via ideological power; nor does it deal sufficiently with social power as it manifests in gender relations and relations within extended families and neighbourhoods (Enarson and Chakabarti 2009; Enarson and Morrow 1998). PAR does acknowledge local conflicts of interest, but these are understood as cases of the exercise of economic power, an example being landlordism. This all implies the need for at least a partial autocritique. This is not the place for that; nor would I abandon a structuralist, political-economic account of vulnerability. I simply want to complement it with analysis that is appropriate for the lived experience of risk.

In what follows, I take the standpoint of critical political ecology (CPE) (Wisner 2019a). So far, CPE has seldom included phenomenological treatment of individual experience as part of the continuum of nested scales where power is manifest in human–environment relations, leaving a crucial gap as I have mentioned in the case of the PAR framework. I define political ecology as a community of research practice using mixed qualitative and quantitative methods to study economic, political, and social power and their influence over how society and nature interact – from the household to the globe, across scales. When engaged as advocates for, and allies of, groups oppressed and exploited by power relations, critical political ecology (CPE) stands with these subaltern groups and their social movements, using its tools to clarify practice and to promote change (Pellow 2018; Wisner 2019b).

Ignoring the life world

By life world I refer to a concept that is basic to the work of Edmund Husserl and Martin Heidegger, two German philosophers who attempted to ground the human sciences in the immediacy of existence (as opposed to the classical search for an 'essence' of objects and human experience). The concept was further elaborated by phenomenological sociologist Alfred Schutz. For all three of them (Britannica 2019, online):

> Life world (*Lebenswelt*) … refers to the world as immediately or directly experienced in the subjectivity of everyday life, as sharply distinguished

> from the objective 'worlds' of the sciences, which employ the methods of the mathematical sciences of nature; although these sciences originate in the life world, they are not those of everyday life.

In the context of risk, lifeworld is the setting for dynamically changing, shifting experiences of safety, belonging, being-at-home and threat, otherness, and displacement (Schutz and Luckmann 1973; Buttimer 1976; Tuan 1976; Buttimer and Seamon 1980). These experiences as well as their concrete, immediate qualities are influenced by a kaleidoscopic mixture of cultural, psychological, social, economic, and environmental stresses and satisfactions. In this context, one can see a link between the 'outside' view of the PAR framework and the 'inner' experience of risk (Tuan 1979). The exercise of economic and political power and coercion influence the subjectivity of everyday life, as suggested by three examples, from Somalia, Chile, and Bangladesh.

Somalia

For residents of Somalia in the 1990s, risk took the form of livestock raiding and the proximity of warlords, drought, and fluctuation in the price of exported camels and other livestock on the side of stress (essentially, the ratio of livestock and grain prices); whilst satisfactions took the form of being surrounded by extended family and kin, good rains and affordable terms of trade for sale of livestock and purchase of grain (Wisner 1994, 1995). These are the experiences as well as material realities that constitute the largely unconscious ground of what conventional disaster studies would call 'risk perception'. However, the experiences associated with material realities are dynamic and dependent on the proximal exercise of social and coercive power and the more remote exercise of coercive, economic, administrative, and political power. In the Somali case, in the early to mid-1990s, kinship relations involved obligations and rights enforced by social power; local warlords may have demanded tribute or conscripted young men as fighters. Warlords may also have provided protection and food (often diverted from overseas humanitarian aid). Distant government or military advisors may have made demands for information and intelligence or militia service, and even more distant UN agencies, INGOs, and camel traders in the Middle East will have exercised economic power. Distant and nearby exercise of economic, political power, and coercion are echoed in the subjective experiences of seasons and the landscape by these farmer-herders in expressions that are inflected by their culture and language.

Somali households were in the difficult position of juggling and making sense of four largely competing discourses: (a) official Somali government's discourse of national security versus terrorism; (b) clan-based separatist ('war lord') discourse of the corrupt, illegitimate state based in Mogadishu; (c) clan and extended family discourse of inward-focused solidarity and self-reliance; (d) received Islamic discourse of the brotherhood of Islamic believers and the injunction to give support to the needy, albeit 'the other' (e.g. people belonging to other clans). At various times, different elements of all four were in sync whilst others were incompatible with the everyday beliefs and knowledge informed by experiences and practices that make up the life world and specific interpretations of risk.

Chile

In another example of the way life world and its experience interact with external power, the following is how Chilean researcher Manuel Trioni reflected on the view of a post-disaster scene in an informal settlement on a hillside in the Town of Messana, where wildfire had burned (Trioni et al. 2019, online).

> Those [burned] stumps haunted me. Not fully trees anymore, not completely rubble yet, they stood in an uncanny space between life and death. They also messed with linear temporalities. Were they directing affect to a past tense, to an event that already happened or to the current vulnerability of Messana? Or were they an admonition of what could occur in the future, again, if the forestry industry that looms [over] Messana and many informal settlements across Chile, is not stopped from converting eucalyptuses, the very same brutally carbonised eucalyptuses now standing in front of me, into flammable material ready for climate-induced mass combustion?

We hear the author interpret his experience of the 'uncanny', of being 'haunted' in the context of the external economic power of the forest industry and the political challenge facing people living in 'many informal settlements across Chile' to get government and industry to heed the 'admonition'. Words such as 'looms' and 'brutally' give the reader additional clues to the deep emotional impact of this scene on the writer, who is both a scientist who thinks in terms of 'linear temporalities' and an ordinary human being viewing 'brutally carbonised eucalyptuses'. Power relations permeate this very personal testimony. The testimony points toward a counter-hegemonic discourse opposed to the Chilean state's story of modernisation and the rationality and benefit of the market. Unlike the Somali herder, here the author, Trioni, does not balance or attempt to integrate incompatible beliefs and conceptions of risk. His science as well as his lived experience come down squarely on a view opposed to the dominant discourse.

Bangladesh

Sultana (2011) explored the challenges faced by Bangladeshi women in providing water free of arsenic for their children to drink. During the 1970s and 1980s some ten million boreholes were drilled by the government, international aid organisations, and individuals without adequate understanding of the underlying geology. Sultanas reveals the tactical and strategic behaviour of these women and the women's emotional experience of negotiating the intersection of economic and social power. The water in many of the wells is contaminated with naturally occurring arsenic. Some wells are safe. Some are not. They are now identified by red and green paint. Each well is on land owned by someone in a village. Access to safe water requires knowledge of family histories and one's economic and social status. The result is a variety of different strategies including gift-giving or barter, purchase, appeals to ethical norms, and even begging or theft. Women reported feeling many emotions such as anxiety and humiliation in the course of obtaining water from unsanctioned safe sources – sometimes at night.

These many emotions reported and the wide variety of local power relations the women have to negotiate yield a variegated pattern of discourses. In the background is the culture of international 'development' and Bangladeshi state discourse of 'modernity' and national development that made the drilling of boreholes possible in the first place. Perhaps some women with exposure to campaigning Bangladeshi NGOs are aware of the culpability of the British Geological Survey and UNICEF for not testing for naturally-occurring arsenic (Atkins et al. 2006). Most of the women interviewed by Sultana weave their appeals to neighbours out of elements of several discourses: (a) shared motherhood; (b) charity as prescribed by Islam; (c) clan and extended family solidarity; (d) and material interest to the extent that capitalist market conditions and ways of thinking have penetrated the locality. Within the context of the daily search for safe water, these discourses provide the background for life worlds containing a number of risks besides exposure to arsenic: (a) other water-born, water-washed, and water-related risks (diarrhoea, scabies and trachoma, and malaria, etc. (White et al. 1972, 2002); (b) humiliation; (c) physical violence; (d) sexual abuse.

Intermezzo: the role of language and performance

The subtle role of discourse may be seen in the seduction of minds and manipulation of life worlds. The role of discourse can be seen in these three cases in which life world interacts with economic power, social power, and coercion. To take a very simple example, the term 'natural disaster' remains current in popular media and everyday speech despite the fact that the scientific and policy worlds now accept that there is nothing 'natural' about a natural phenomenon becoming a disaster for human beings, and that a variety of human actions or failures to act turn a natural event into a human disaster (O'Keefe et al. 1976; Gould et al. 2016). 'Common sense' tells most non-specialist, lay people that earthquakes, hurricanes, and floods are responsible for disaster. That belief is part of 'common sense', defined by Giddens (1984, 337) as 'propositional beliefs implicated in the conduct of day-to-day activities'. It is also an instance of what Gramsci calls 'hegemony' (Gramsci 1971; García-López et al. 2017).

The state uses ideological and linguistic hegemony to encourage beliefs that underpin ideological stability of the state itself and the dominant economic system. Both depend on the popular perception of these institutions as benign if not directly helpful, desirable, or even necessary. To interpret one's experience of disaster as caused by malfeasance by the state or actions by the owners of capital would undermine confidence in these institutions. Individuals are further rendered docile and less questioning of the state and economic system by feelings of awe in the face of professional knowledge and dependency on experts to educate their children, protect their health, and to protect them from 'natural disasters' (Freire 1970; Illich 1982, 1985, 2000).

Manipulating the life world

Individuals belong to groups with cultures. Local knowledge is absorbed into cultural practices and diffused throughout a group of people sharing language, history, and location – integrated as the experience of place – as well as being transmitted to following generations. When local government or other outside agencies such as international non-governmental organisations involve local people for purposes of 'risk communication', the outsiders often use the term 'participation'. However, involvement of local residents and risk bearers in planning and decision-making can take many forms, some involving quite superficial engagement and others much greater (Arnstein 1969; Cooke and Kothari 2001).

Government officials or INGO staff may redeploy local knowledge in order to manipulate consensus, acceptance of a pre-formulated government 'solution' or to enforce normalised 'good' disaster risk behaviour. Local knowledge that arises out of concrete experience of life world and the material world – say vernacular terminology for weather phenomena or soil types – may simply be catalogued by external experts for purposes of better translating their own preconceived disaster risk messages (Gibson and Wisner 2016). Sillitoe and Marzano (2009, 16) observe that development agencies want 'to give indigenous knowledge coherence and structure' so it can be applied generally to solve problems like soil fertility. However, they warn against the ethnocentrism entailed by applying an overarching theoretical framework within which local knowledge systems are *made* to fit.

Consider three examples in which such a top-down drive for coherence acknowledges but bypasses or directly dismisses local life worlds or seeks to manipulate them for preconceived ends.

Botswana

Commenting on community-based resource management programs in Botswana, Blaikie (2006, 1953) concludes that these may turn out to be Trojan Horses that provide opportunities for new local and national entrepreneurs to exert control over natural resources. Community-based, participatory, decentralised management involves the creation of institutions and what Blaikie calls 'blue printing' of the process, or 'designing the local so as to render it manageable'. The result is suppression of diversity. The diversity of meanings suggested by local life worlds finds no avenue for expression in the formal process of managing woodland resources. A woodlot can have many meanings depending on whether it is 'a sacred grove, a supply of fuelwood, a bio-diverse collection of medicinal plants, high quality carving wood for tourist curios, or acts as protection for a watershed' (Blaikie 2006, 1952).

Diversity of power and access among the people represented by the nominally 'community-based' process is also suppressed by the institutional demand for standardisation and replicability. Blaikie describes the 'black-boxing' of 'troubling complexities' such as differences in wealth and political access among households, and among men, women, children, and ethnic minorities. In Botswana, the Basarwa minority was told by government officials, 'we are all Batswana now' (citizens of Botswana). They were told not to try to preserve their cultural identity. This identity is inextricably bound up with hunting and the life world that provides the detailed local knowledge and skill for hunting and using and sharing the product of the hunt. When the Basarwa protested erosion of their hunting rights, they were told they had to agree to the creation of a Community-Based Natural Resource Management (CBMRM) trust, or the government would create it without them, and the Basarwa would lose out (Blaikie 2006, 1953). Stone and Nyaupane (2014, 17) describe 'hasty clustering' of beneficiaries and found that 'a single CBMRM implementation model approach fail[s] to incorporate local variations in natural resources, culture, and socioeconomic conditions'. The result is the same whether this is an instance of the conscious use of participation as a Trojan Horse for the benefit of private ranchers or an instance of the deafness of the state.

Nepal

Similar 'blue printing' and 'hasty clustering' were present in the Government of Nepal's nation-wide programme to create 'disaster-resilient communities' (Flagship 4 2013; Ovens et al. 2016; Panley 2019). The initiative also involved contact between Western scientific hydrology and national hierarchical forms of governance with indigenous life worlds, institutions, and water knowledge. The Nepal government's efforts to create community disaster resilience were based on convincing citizens to join village-scale groups that would receive training in early warning, preparedness, and response. These primary groups were to be supported by a hierarchy of committees and scores of paper plans at the national, provincial, district, and rural municipality scale. Given the nature of hierarchical bureaucracy, most of the training for the village disaster committees involved top-down instruction about earthquake, landslides, and floods, with little attention paid to the local people's perceptions and cultural interpretations of aspects of the natural surroundings, including what outside scientists define as 'hazards' (Ovens et al. 2016).

This same top-down approach had been applied in eastern Nepal for the development of a warning system for glacial lake outbreak floods (GLOFs) (Kattlemann 2003). However, in that instance citizens' motivation was low and buy-in temporary and limited (Paul et al. 2018). Residents had been used as 'citizen sensors' and inexpensive labourers. They were not involved in every stage of the design of the warming system; neither had there been a dialogue between lay understanding based on their life world (beliefs, knowledge tacit in daily activities, attachment to place). On the other hand, yet another project in western Nepal that built on local groups succeeded in engaging the interest and support of the village because it involved citizens in every stage of installing sensors, monitoring and interpreting rainfall and landslide hazard (Malakar 2014). Interpretation involved engaging with the differences in hazard perception between the culture of a universal science and Nepali local science (Sillitoe and Marzano 2009).

Peru

The story of misunderstanding and manipulation of local life worlds seen in Botswana and Nepal has been repeated in the highlands of Peru, perhaps in an even more extreme form. Andean Peru has seen numerous landslides and been the locale for many studies and projects focusing on landslide hazard mapping and warning. It is also a terrain within which formerly isolated peasant farmer settlements have strong cultural beliefs, attachment to the land as well as recent and historical reasons for suspicion and mistrust of outsiders. Many of the external interventions had been technical and were not just ignorant of local understandings and coping mechanisms, but disrespectful of them (Klimes et al. 2019) whilst attempting to take advantage of local institutions and local labour power for implementation and maintenance of new infrastructure and instrumentation (Maskrey 2011).

In the case of Rampac Grande in Peru's central Andes, both this top-down approach and a more successful full integration of the community and dialogue between outside and inside knowledges took place over a period of nine years. Klimes et al. (2019) narrate the long process that began with a deadly landslide in this small peasant community in 2009. There followed a series of external technical assessments the result of which were never communicated to the residents. Meanwhile, the Quechua-speaking locals decided that the landslide was likely triggered by explosions set off by illegal miners. There was a history of disputes with foreign-owned mines in the region. They decided in their local government institution (*asemblea*) to close the community off from outsiders since they were dangerous, and their studies did them no good. However, in 2014 the community once again admitted a research team to do hazard mapping with the hope that it might lead to government investment. There followed a prolonged confidence-building process, mapping, and installation of an extensometer to record the pattern of land surface slippage. All community institutions and the *asemblea* were involved at all stages in the project, and this seems to have converted the villagers' earlier mistrust and resistance into full ownership.

An important part of confidence building, and mapping involved group discussions to find Quechua vocabulary for the processes that were being mapped.

It was found that some local understandings and outside views could coincide if appropriate language made communication possible. In other cases, the two sorts of knowledge conflicted, but could be allowed to co-exist. Importantly, the local view was that earth beings such as the mountain were personified and believed to have a role and agency in the processes and decisions underway. It was believed that the 'mountain had to rest and heal' after the landslide before pre-slide land use such as grazing the affected slope and settlement could continue. The act of populating political space with spirit and earth beings is a well-known aspect of 'indigenous cosmopolitics' in the Andean countries (de la Cadena 2010). On this basis, affected households were resettled locally. Such a view is fully compatible with an external standard approach to post-disaster practice. The evident willingness of the research team to engage with villagers concerning their life world and beliefs was an essential part of winning their acceptance.

Attacking the life world

Rodrigo Mena spent many months interviewing humanitarian responders and even longer in developing a broader understanding of disasters that take place during high-intensity conflict in South Sudan, Afghanistan, and Yemen (Mena 2020). He summarises the conversation in meetings of these humanitarian staff. His observations provide a segue to my next cluster of reflections. Mena suggests that individual people's local knowledge and skill can simply be abandoned, written off.

> In meetings… the most recurrent question when performing triage was whether it was feasible to respond, considering the insecurity in the country, access constraints and the lack of infrastructure and services… A participant in one of these meetings later told the [me] that 'Making the decision not to help everyone is painful'… For many research participants … the lack of funds was one of the primary factors hindering humanitarian aid and forcing them to triage among the multiple affected groups and places. … Local capacities and community resilience were another aspect playing a dominant role in meetings. …[A]ffected populations develop multiple coping and survival mechanisms. The problematic part of this in South Sudan, however, is that the conflict has damaged many of those coping mechanisms. … Although the research participants often realized that the resilience/capacity argument may be unjustified and mask a more complex and changing situation, … the routine use of this argument may be seen as another means for aid workers to alleviate the 'pain of triage'.
>
> (Mena 2020, 67–68)

Humanitarian organisations have power in relation to people caught up in disaster and conflict; however, these humanitarian organisations are subservient to institutions and actors who have much more power and whose decisions about the use of public and private wealth, public policy, and investment determine vulnerability to natural hazards and the options civilians have when threatened by conflict and natural hazards.

Austerity and precarity

At the scale of national political economies, we live in an era of accelerating globalisation, erosion of the economic rights and welfare systems that prevailed from the end of WWII until the 1980s, and growing income and wealth disparity (Piketty 2017). Under these conditions, power manifests itself in austerity policies and the precarity of the marginally employed. *A general definition of austerity* is '[d]ifficult economic conditions created by government measures to reduce public expenditure' (Lexico 2019 [2015] cited in The Economist 2015). This generally involves cutting public sector employment, reducing expenditure on education, health care, and other social services, and often neglect of public infrastructure, and even retrenchment of pensions and other social benefits. There were rounds of such austerity imposed on African and Latin American governments in the 1980s and 1990s, on Greece during the Euro Crisis in the 2000s, in Great Britain and France, and it was imposed on the heavily indebted government of Puerto Rico both before and after Hurricane Maria. Naomi Klein writes (2017, 253)

> [b]efore those fierce winds came, the debt (illegitimate and much of it illegal) was the excuse used to ram through a brutal program of economic suffering, what the great Argentine author Rudolfo Walsh, writing about four decades earlier, famously called *miseria planificada*, 'planned misery'.

Precarity is defined as follows by Kasmir (2018, on line):

> Precarity emerged as a central concern in scholarly research and writing in the twenty-first century, partly in response to political mobilizations against unemployment and social exclusion. Together with related concepts – such as precarious, precariousness, precaritization and 'the precariat' – precarity refers to the fact that much of the world's population lacks stable work and steady incomes. Informal, temporary, or contingent work is the predominant mode of livelihood in the contemporary world, where garbage picking, performing day labor, selling petty commodities, and sourcing task-based 'gigs' through digital platforms exemplify some of precarity's many forms.

People at the sharp end of such national-scale processes in much of Europe and North America, and with different configurations elsewhere, cannot count on the state to assure them enjoyment of two of US President Roosevelt's famous Four Freedoms: freedom from want and freedom from fear. The huge material losses in the US from Hurricanes Florence (2018) that affected the southeastern states, Hurricane Michael (2018) that hit the Florida panhandle and Hurricane

Katrina (2004) in the Gulf States accelerated precarity and were signals of under-investment by governments in risk reduction. Low-income, insecurely-employed people, especially racial minorities, lost the little wealth they and their parents and grandparents had accumulated, that is: their homes. Many were uninsured. Their livelihoods were interrupted or destroyed because many of their small business employers did not reopen. Recovery was a painfully slow process.

A third of New Orleans' pre-Katrina population never returned to the Ninth Ward and other low-lying, low-income neighbourhoods (Bates and Green 2009). Over 50 per cent of the residences in the predominantly African-American Lower Ninth Ward were not receiving mail in 2014, nine years after Hurricane Katrina (Adelson 2015). Some of the low-income African-American population was evacuated to Houston and have stayed, settling in low-income, areas of East Houston (Jan and Martin 2017). Those areas of Houston were later flooded in 2017 by Hurricane Harvey's torrential rainwater draining away from White sub-urban sprawl in the Northwest (Kaufman 2017). As the earlier citation by Klein suggested, Puerto Rico offers another example of a hazard, Hurricane Maria, impacting a situation of extreme marginality and pre-existing insecurity in 2018 (García-López 2018).

Displacement and risk creation

Power also asserts itself by displacement of people to make room for speculative investment in cities and in extractive industries and infrastructural megaprojects in rural areas. Such investments often create risk by changing the material environment in situ, directly or indirectly causing people to shift to new locations that often contain unfamiliar hazards, by undermining or rendering impossible livelihood activities, whilst also undermining place-based identity, knowledge, and, hence, the life world (Vickery 2018).

The life world is made up of the many details and practices of daily life that constitute dwelling in a place. Life world and place are not identical because individuals bring unique sets of interpersonal relationships, memories, dreams and expectations to the process of dwelling, or place making (Cresswell 2015). Nevertheless, forced uprooting from place seriously challenges and rearranges a person's life world. It takes time for a person to form an attachment and understanding of a new place. During this transition period, new hazards may not be understood, and the displaced newcomers may be more vulnerable to them.

At the extreme, such dislocations and displacements can leave people in a condition that Agamben refers to as 'bare life' (Agamben 1995). This is the existence not of a citizen or human as political and historical agent, but as an organism whose basic needs must be met to stay alive. He writes, 'The fundamental categorical pair of Western politics is not that of friend/enemy but that of bare life/political existence… exclusion/inclusion' (Agamben 1995, 8). Agamben's notion of bare life is clearly applicable to people marginal to the dominant political and economic system. Agamben originally applied the idea of bare life to concentration camps. One thinks of other exceptional situations such as prisons and the immigrant detainee camps and installations on the US-Mexican border or in Libya. Historically, there was the workhouse (Kearns 2007). However, even in the case of less strictly controlled refugee camps, people are already learning about the new life world into which they have been suddenly thrust. In such places people attempt to supplement the conditions of bare life – not just an act of coping but one of resistance.

Whereas Agamben understands bare life as a condition imposed by the state on populations, I believe there is more to the story. There is usually also resistance and push-back (Oliver-Smith 2010; Scott 1985). Nevertheless, despite resistance, the state remains hugely influential in issues of public health (Packard 2016), social policy and food security. In general terms, the state has defined the 'deserving' and 'undeserving' poor (Seabrook 2014), and often the state has explained neglect or scanty provision for suffering people by applying stigma: the nineteenth-century Irish catholic peasantry was impossibly 'lazy'. One might compare the othering of African small farmers today, who are labelled 'non-progressive' by bureaucrats who wish to welcome international agribusiness companies into vast areas of arable land in order to 'modernise' (Smucker et al. 2015).

Whilst the state has a role; my argument turns on two additional moments that nuance the top-down determinism implied by the notion of bare life and more generally in writings about biopolitics (Foucault 1991). Firstly, there is the dialectical moment of resistance and efforts to cope such as the place-making work of Somali women and solidarity and mutual aid among them (see Box 10.1). Secondly, just as the logic of capital accumulation drives groups of people 'towards the edge' *metaphorically* in the sense of bare life conditions, so also this market logic may *literally* push flows of migrants to the margins and extremes of a territory where they face hazards. This is certainly what I found in my Ph.D. work on the small farmers in Kenya's arid and semi-arid lands in the 1970s. Kenyan independence had changed land ownership, land use, and marketing relations in the highland core so that a class of Black Kenyan coffee farmers and other entrepreneurs consolidated ownership. Poorer farmers in debt lost land and migrated to the less fertile and less well-watered periphery. An expanded core produced and expanded periphery, and the process continued through the 1980s and 1990s (Wisner 1978; Smucker and Wisner 2008).

Others and I have used the term 'marginalisation' where today bare life and precarity may be used (Wisner 2010; Blaikie and Brookfield 1987). In my early Kenya work, I referred to relatively isolated rural people, geographically remote from centres of power and decision making, with few economic resources and access to only scanty or depleted natural resources (thus the phase 'eco-demographic and political marginality' in Wisner (1978). However, marginalisation resulting in bare life and extremes of vulnerability to hazards are not only to be found in peripheral zones of low-to-medium income countries. They also exist in high-income countries (Vickery 2018), and in urban regions of medium and low-income countries.

BOX 10.1 WOMEN'S PLACE MAKING AND RESISTANCE IN SOMALIA

In 1981, I carried out interviews with refugees in a camp in western Somalia near the Weli Shabelle River (Lewis and Wisner 1981). The camp was for ethnic Somalis, 90% of them women and children, who had fled fighting in the Ogaden region of Ethiopia that contained a minority population of ethnic Somalis. After only a year, women had constructed a life world together much different than the world that the Somali military commander of the camp sought to impose. Whilst newcomers were supposed to have been allocated randomly into sections of this camp of 40,000 souls, the women re-organised themselves by clan and extended family. Women learned to trade camp provisions such as cooking oil for soap and other things they desired with traders from the nearby town, Beledwayne. Woman set up small businesses such as a bakery and beauty salon. They enhanced the structure of their beehive-shaped tents and even attempted to beautify them. The women took turns at child care while others would stand in line for the various food and water distributions. Weaving elements of prior spatial, social, and economic relations with new ones, these women seemed to have been constructing a new life world that provided marginally more materially and also offered them a socio-psychological anchor and meaning. Nevertheless, this was a desperate situation, and one should not romanticise the solidarity of the Somali women or their creative coping and place-making. People have limits, and in this new environment flooding was one of the new physical hazards the women in the Somali refugee camps had to learn to cope with, as was sexual predation by camp authorities.

The case of Zika virus in Brazil provides another example. Viewed in the context of the political ecology of health and disease in Brazil, the causal cascade or progression of vulnerability is clear. As Brazilian law professor, Debora Diniz, explained (Diniz 2016, online)

> The epidemic mirrors the social inequality of Brazilian society. It is concentrated among young, poor, black and brown women, a vast majority of them living in the country's least-developed regions. The women at greatest risk of contracting Zika live in places where the mosquito is part of their everyday lives, where mosquito-borne diseases like dengue and chikungunya were already endemic. They live in substandard, crowded housing in neighborhoods where stagnant water, the breeding ground for disease-carrying mosquitoes, is everywhere. These women can't avoid bites: They need to be outdoors from dawn until

> dusk to work, shop and take care of their children. And they are the same women who have the least access to sexual and reproductive health care.

'Modern' economic exploitation of rainforests in several parts of the world also provides vivid examples of the destruction of life worlds and livelihoods. In Brazil, Indonesia, Cameroon indigenous forest dwellers (IFDs) are being displaced from ancestral land by policies that promote forest clearance for mechanised farming and grazing, logging, and mining. Land use by IFDs is said to be pre-modern and therefore inefficient, despite studies that show that IFDs use forest resources to produce livelihoods efficiently and sustainably (Barham et al. 1999), whilst also maintaining biodiversity (Sayer et al. 2012).

Brazil's Belo Monte hydroelectric project is an example of the intrusion of megaprojects into the Amazon that predates Brazilian president Bolsanaro's renewed push to colonise it. Dams on the Xingu River began to fill in 2015 and caused flooding that eventually displaced tens of thousands of people living in and from the forest, including many Kayapó IFDs. Changes in the river's regime and ecology have destroyed fishing livelihoods, whilst the large lake produced has excluded IFDs from access to forest resources. Some of these displaced people have settled on the periphery of the city of Alta Mira, where the dam's effect on the river has flooded some of the city's low-income neighbourhoods (Marchezini and Wisner 2017; Anderson and Elkaim 2018).

In Cameroon, IFDs were displaced when the government leased large areas of its southern rainforest to rubber and timber companies (Rights and Resources 2013). The displaced include approximately 5,000 people from five villages in the South Region, as well as the Bagyeli pygmy community. They were offered settlement on other land along a road on the margins of the development area; however, they cannot continue hunting from that location as the core of their livelihood system. In addition, the land offered and planting practices required are unfamiliar and do not have meaning within their customary life world.

This process has continued. By 2019, 17.5 million hectares of the country's 22.5 million hectares of forested land had been classified as 'productive' and allocated to logging companies and other private companies (Mohanty 2019). The Chad-Cameroon Pipeline project alone will affect many thousands of IFDs (CED et al. 2010). People have been evicted without compensation when the government rejects their claim of ancestral occupancy. Only ten per cent of the IFDs have access to tribal hunting grounds even though 60% of the Bakosi National Park is classified as a 'community forest area' (Mohanty 2019). In April 2019, the African Commission on Human and Peoples' Rights sent a letter to the government of Cameroon expressing concern about the treatment of indigenous people (ISHR 2019).

The case of Indonesia is depressingly similar and perhaps worse. Violence has been used to evict IFDs from their villages so that palm oil production can be expanded. In 2013, armed men and bulldozers invaded four villages of Suku Anak Dalam people in Jambi Province in Sumatra, looting and destroying houses. The beneficiaries of expanded land for production are companies such as a majority Finnish-owned corporation that refines palm oil into biodiesel in Singapore and Rotterdam and other companies (Rainforest Rescue 2013). By 2019 the European Commission began to take steps to phase out palm oil for biodiesel as part of Europe's energy strategy. That year Indonesian police were investigating two dozen companies accused of starting fires in order to clear new land for oil palms (Reuters 2019). The incident in Jambi Province was not unique. In 2016, the Orang Rimba people were attacked, their possessions and homes burned as part of their eviction from ancestral land wanted for palm oil plantation expansion (Survival International 2016). Eviction, degradation of the basis of livelihoods and relocation with inadequate and corruptly-administered compensation has affected many of Indonesia's 2,320 indigenous groups comprising about a quarter of the national population (Human Rights Watch 2019).

In these cases, one sees that political and economic power not only exposes people to known risks but can also create new ones. Investments such as mega-projects that involve large-scale reshaping of urban and rural space throw the life world into disarray. Disaster risk creation (DRC) is the evil twin of 'disaster risk reduction' (DRR), the upbeat buzzword dear to international organisations, governments and INGOs. Some earlier authors have used this rather shocking turn of phrase when discussing the historical and structural (or 'root') causes of people's vulnerability to the impacts of natural hazards (Lewis and Kelman 2012; Jerolleman 2019). The displaced must learn to negotiate new environments and new threats, and they leave behind some people who live in altered environments and face new hazards. Dislocation of this kind not only exposes displaced people to new hazards but rips them from familiar life worlds and demands they learn to understand and cope with hazards in a new environment.

Conclusion: a better normal?

When students of disaster take power seriously, they are able to contribute to a critical assessment of policies that disregard, condone, or are complicit in displacement and risk creation. Disaster studies can also help to understand increasing citizen resistance to the systems of power that have brought the planet and humanity face to face with increasing daily risk for some, existential risk for all. Ulrich Beck believed that confronting global risks such as climate change, terrorism, pandemic disease, and financial crisis will provoke a radical transformation among the stakeholders within the modern nation state. He wrote in 2011 (2011, 15–16):

> Action strategies, which global risks open up, overthrow the order of power, which has formed in the neoliberal capital-state coalition: global risks empower states and civil society movements because they reveal new sources of legitimation and possibilities of action for these groups of actors; on the other hand, they disempower globalized capital because the consequences of investment decisions and externalizing risks in financial markets contribute to creating global risks…

However, one needs to go further than Beck. His invocation of 'empowerment' of civil society does not include tapping the anger and grief of the displaced and those living a precarious or even bare life. Attacks on the lifeworld and the creation of risk are capable of producing such strong emotions and also motivating demands for change. Should one extend and specify Beck's generalisation in this way? Waite (2009) would answer 'yes'. Her review of the history of the term precarity and the possibility of a critical geography of precarity, reveals the dual character of the term: both a condition and site for mobilisation against the structures of power that produce precarity and vulnerability. Another voice answering 'yes' is Garriga-López (2019), who studied the experience of abandonment by the US federal state and government of Puerto Rico following Hurricane Maria and the rise of mutual aid networks and local initiatives in health care, solar energy and food security. But then there are other questions: Who speaks for the precariat? How can their own voices be set free, and their own passion to reclaim and to rebuild their own place in the world?

What is the position of the researcher in all this? Part of the change that must take place is the emergence of the activist or engaged scholar, who stands with, and is an ally of the marginalised, displaced and those made to live precarious lives by the systems revealed by research (Wisner et al. 2005; Wisner 2019b). A dialogical, decentralised, activist approach requires engagement with people who are risk bearers and with organisations and networks of the displaced, excluded, and marginalised. Examples of such organisations are the branches of Slum Dwellers International http://knowyourcity.info/, Disability Rights International https://www.driadvocacy.org/, Huairou Commission https://huairou.org/, and Via Campesina https://viacampesina.org/en/. Also, there is the Global Network of Civil Society for Disaster Reduction (https://www.gndr.org/), which includes many platforms that give voice to marginalised and excluded people in its 2019 *Views from the Frontline* survey (GNDR 2019). The survey involved a conversation among people about their own capacities in the face of change and threats in more than 50 medium and low-income countries. The data is to be used simultaneously for international advocacy, commentary on national policy, and also for local action planning. The intention is to use the locally-based process to raise consciousness and begin a process of social and political mobilisation. The anticipated results include various local demands on the state and on corporations for transformation (Pelling et al. 2015), or, in some cases, the circumvention of the state and market as groups of people create their own alternatives (Gibson and Wisner 2016).

References

Adelson, Jeff. 2015. 'Hurricane Katrina transformed New Orleans: The region's makeup after unrivaled exodus in the US'. *The Advocate*, 23 August https://www.theadvocate.com/baton_rouge/news/article_d1bd4e2f-396b-5559-ad2a-baa37968d45e.html (see map of residences receiving mail in 2014 by U.S. Census Bureau Data Center analysis of Valassis Residential and Business Data Base).

Agamben, Giorgio. 1995. *Homo Sacer: Sovereign Power and Bare Life*. Stanford, CA: Stanford University Press.

Anderson, Maximo and Elkaim, Aaron. 2018. 'Belo Monte legacy: Harm from Amazon dam didn't end with construction'. *Mongabay*, 23 February https://news.mongabay.com/2018/02/belo-monte-legacy-harm-from-amazon-dam-didnt-end-with-construction/

Arnstein, Sherry. 1969. 'A ladder of citizen participation'. *Journal of the American Planning Association* 35 no. 4: 216–224.

Atkins, P., Hassan, M., and Dunn, C. 2006. 'Toxic torts: Arsenic poisoning in Bangladesh and the legal geographies of responsibility'. *Transactions of the Institute of British Geographers* 31 no. 3: 272–285.

Barham, B., Coomes, O., and Takasaki, Y. 1999. 'Rainforest livelihoods'. *Unasylva* 50 no. 198: 34–42.

Bates, Lisa and Green, Rebekah. 2009. 'Housing recovery in the Ninth Ward'. In: *Race, Place, and Environmental Justice after Hurricane Katrina*, edited by Robert Bullard and Beverly Wright, 229–248. Oxford, GB: Routledge.

Beck, Ulrich. 2011. 'Living in and coping with world risk society'. In: *Coping with Global Environmental Change, Disaster and Security* edited by Hans-Gunter Brauch et al., 11–17. Berlin: Springer.

Blaikie, Piers. 2006. 'Is small always beautiful? Community-based natural resource management in Malawi and Botswana'. *World Development* 34 no. 11: 1942–1957.

Blaikie, Piers and Brookfield, Harold. 1987. *Land Degradation and Society*. London: Longman.

Blaikie, Piers, Cannon, Terry, Davis, Ian, and Wisner, Ben. 1994. *At Risk: Natural Hazards, People's Vulnerability and Disasters*. London: Routledge.

Britannica, Encyclopedia. 2019. 'Lifeworld'. https://www.britannica.com/topic/life-world.

Buttimer, Anne. 1976. 'Grasping the dynamism of the life world'. *Annals, Association of American Geographers* 66: 277–292.

Buttimer, Anne and Seamon, David. 1980. *The Human Experience of Space and Place*. London: Croom Helm.

CED (Center for Environment and Development), RACOPY (Reseau Recherches Actions Concertee Pygmees and FPP (Forest Peoples Program). 2010. *Indigenous Life World Peoples' Rights in Cameroon*. Supplementary report submitted in connection with Cameroon's second periodic report. Submitted to the African Commission eron Human and People's Rights.

Cooke, Brian and Kothari, Umma. 2001. *Participation: The New Tyranny?* London: Zed.

Cresswell, Tom. 2015. *Place: An Introduction*. 2nd edition. London: Wiley-Blackwell.

de la Cadena, Marisol. 2010. 'Indigenous cosmopolitics in the Andes: Reflections beyond 'politics''. *Cultural Anthropology* 25 no. 2: 334–370.

Diniz, Debora. 2016. 'The Zika virus and Brazilian women's right to choose'. *New York Times*, 8 February.

Enarson, Elaine and Chakabarti, Dhar, eds. 2009. *Women, Gender and Disaster: Global Issuesand Initiatives*. Beverly Hills, CA: Sage.

Enarson, Elaine and Morrow, Betty, eds. 1998. *Gendered Terrain of Disaster: Through Women's Eyes*. Westport, Conn.: Praeger.

Flagship 4, Nepal Risk Reduction Consortium. 2013. *Handbook: Nepal's 9 Minimum Characteristics of a Disaster Resilience Community*. Kathmandu: Ministry of Federal and Local Development and the International Federation of Red Cross and Red Crescent Societies http://flagship4.nrrc.org.np/document/flagship-4-handbook-english.

Foucault, Michel. 1991. *Discipline and Punish*. London: Penguin.

Freire, Paulo. 1970. *Pedagogy of the Oppressed*. New York: Continuum.

García-López, Gustavo. 2018. 'The multiple layers of environmental injustice in contexts of (un)natural disasters: The case of Puerto Rico post-Hurricane Maria'. *Environmental Justice* 11 no. 3. Published Online: 01 June. doi:10.1089/env.2017.0045.

García-López, Gustavo, Velicu, Irina, and D'Alisa, Giacomo. 2017. 'Performing counter-hegemonic common(s) senses: Rearticulating democracy, community and forests in Puerto Rico'. *Capitalism Nature Socialism* 28 no. 3: 88–107.

Garriga-López, Adriana. 2019. 'Puerto Rico: The future in question'. *Shima*, September https://www.researchgate.net/publication/336692535_Puerto_Rico_The_Future_In_Question

Gibson, Terry and Wisner, Ben. 2016. ''Let's talk about you': Opening space for local experience, action and learning in disaster risk reduction'. *Disaster Prevention and Management* 25 no. 5: 664–684, doi:10.1108/DPM-06-2016-0119.

Giddens, Anthony. 1984. *The Constitution of Society*. Berkeley, CA: University of California Press

GNDR (Global Network of Civil Society Organisations for Disaster Reduction). 2019. https://www.gndr.org/programmes/views-from-the-frontline/vfl-2019.html

Gould, Kevin, Garcia, Magdalena, and Remes, Jacob. 2016. 'Beyond 'natural-disasters-are-not-natural': The work of state and nature after the 2010 earthquake in Chile'. *Journal of Political Ecology* 23 no. 1: 93–114.

Gramsci, Antonio. 1971. *Selections* from *Prison Notebooks*. London: Lawrence and Wishart.

Human Rights Watch. 2019. ''When we lost the forest, we lost everything': Oil palm plantations and rights violations in Indonesia'. *Human Rights Watch*, 22 September https://www.hrw.org/report/2019/09/22/when-we-lost-forest-we-lost-everything/oil-palm-plantations-and-rights-violations.

Illich, Ivan. 1982. (1974) *Medical Nemesis: The Expropriation of Health*. New York: Pantheon.

Illich, Ivan. 1985. (1973) *Tools for Conviviality*. New York: Marion Boyars.

Illich, Ivan. 2000. (1970) *Deschooling Society*. New York: Marion Boyars.

ISHR (International Service for Human Rights). 2019. 'Cameroon must collaborate with regional mechanisms to guarantee the protection of human rights'. *International Service for Human Rights* 29 May http://www.ishr.ch/news/achpr-64-cameroon-must-collaborate-regional-mechanisms-guarantee-protection-human-rights.

Jan, Tracy and Martin, Brittany. 2017. 'Houston took them in after Katrina: Then Harvey hit'. *The Washington Post*, 29 August. https://www.washingtonpost.com/news/wonk/wp/2017/08/29/houston-took-them-in-after-katrina-then-harvey-hit/.

Jerolleman, Alessandra. 2019. *Disaster Recovery Through the Lens of Justice*. Cham, Switzerland: Springer Nature/Palgrave Macmillan.

Kasmir, Sharryn. 2018. 'Precarity'. In *Cambridge Encyclopedia of Anthropology* edited by Stein, F. Cambridge, GB: Cambridge University Press. https://www.anthroencyclopedia.com/entry/precarity.

Kattelmann, Richard. 2003. 'Glacial lake outburst floods in the Nepal Himalaya: A manageable hazard?' *Natural Hazards* 28: 145–154.

Kaufman, Alexander. 2017. 'Houston flooding always hits poor, non-White neighborhoods hardest'. *Huffpost*, 29 August. https://www.huffpost.com/entry/houston-harvey-environmental-justice_n_59a41c90e4b06d67e3390993.

Klein, Naomi. 2017. *The Battle for Paradise: Puerto Rico Takes on Disaster Capitalists*. London: Zed.

Klimes, Jan, et al. 2019. 'Community participation in landslide risk reduction: A case history from Central Andes, Peru'. *Landslides* 16 no. 9: 1779–1791.

Lewis, Herbert and Wisner, Ben. 1981. *Refugee Rehabilitation in Somalia*. Development Project, Consulting Report No. 6. Madison, WI: University of Wisconsin.

Lewis, James and Kelman, Ilan. 2012. 'The Good, the Bad and the Ugly: Disaster risk reduction (DRR) versus disaster risk creation (DRC)'. *PLOS Current Disasters*. http://currents.plos.org/disasters/index.html%3Fp=1829.html.

Lexico, powered by Oxford. 2019. https://www.lexico.com/en/definition/austerity.

Malakar, Y. 2014. 'Community-based rainfall observation for landslide monitoring in western Nepal'. In *Landslide Science for a Safer Geoenvironment* edited by Kyogi Sassa, Paulo Canuti and Yueping Yin, Vol. 2, 757–763. Cham, Switzerland: Springer.

Marchezini, Victor and Wisner, Ben. 2017. 'Challenges for vulnerability reduction in Brazil: Insights from the PAR framework'. In *Reduction of Vulnerability to Disasters: From Knowledge to Action* edited by V. Marchezini, B. Wisner, L. Londe and S. Saito, 57–96. Sao Carlos, Brazil: RiMa Editores & open access at https://www.preventionweb.net/publications/view/56269.

Maskrey, Andrew. 2011. 'Revisiting community-based disaster risk management'. *Environmental Hazards* 10 no. 1: 42–52.

Mena, Rodrigo. 2020. *Disasters in Conflict: understanding disaster governance, response, and Risk reduction during high-intensity conflict in Afghanistan, South Sudan, and Yemen*. PhD thesis, Erasmus University Rotterdam/International Institute of Social Studies, The Hague.

Mohanty, Abhijit. 2019. 'Tribal communities suffer when evicted in the name of conservation'. *Down to Earth* 10 May https://www.downtoearth.org.in/blog/forests/tribal-communities-suffer-when-evicted-in-the-name-of-conservation-64376.

O'Keefe, Phil, Westgate, Ken, and Wisner, Ben. 1976. 'Taking the 'naturalness' out of 'natural disaster''. *Nature* 260: 566–567.

Oliver-Smith, Anthony. 2010. *Defying Displacement: Grassroots Resistance and the Critique of Development*. Austin, TX: University of Texas Press.

Ovens, Katie, Sigdel, S., Rana, S., Wisner, Ben, Datta, A., Jones, Stuart, and Densmore, Alexander. 2016. *Review of the nine minimum characteristics of a disaster-resilient community in Nepal*. Research Report, Durham University, Durham, UK. https://flagship4.nrrc.org.np/sites/default/files/documents/9mc_final_report_may2017_high_res.pdf.

Packard, Randall. 2016. *A History of Global Health*. Baltimore, MD: Johns Hopkins University Press.

Panley, Chandra. 2019. 'Making Communities Disaster Resilient: Challenges and prospects for community engagement in Nepal. *Disaster Preparedness and Management* 28 no. 1: 106–118.

Parsons, Talcott. 1963. 'On the concept of political power'. *Proceedings of the American Philosophical Society* 107 no. 3: 232–262.

Paul, Jonathan, Buytaert, Wouter, et al. 2018. 'Citizen science for hydrological risk reduction and resilience building'. *Wiley Review – WIREs Water*, 5: e1262. doi:10.1002/wat2.1262.

Pelling, Mark, O'Brien, Karen, and Matya, D. 2015. 'Adaptation and Transformation'. *Climate Change*, doi:10.1007/s10584-014-1303-0.

Pellow, David. 2018. *What is Critical Environmental Justice?* Cambridge, GB: Polity.

Piketty, Thomas. 2017. *Capitalism in the 21st Century*. Cambridge, MA: Belknap/Harvard.

Rainforest Rescue. 2013. 'Indonesia: terror and eviction for palm oil'. *The Ecologist*, 25 `December. https://theecologist.org/2013/dec/25/indonesia-terror-and-eviction-palm-oil.

Reuters. 2019. 'Indonesian police investigate palm oil companies over forest fires: Ministry.' *Reuters*, 29 August. https://www.reuters.com/article/us-indonesia-environment-wildfire/indonesian-police-investigate-palm-oil-companies-over-forest-fires-ministry-idUSKCN1VJ1BS.

Rights and Resources. 2013. 'Cameroon: Forest dwellers lose out as land handed over to developers'. *Rights and Resources* 29 March. https://rightsandresources.org/en/blog/allafrica-cameroon-forest-dwellers-lose-out-as-land-handed-to-developers/

Sayer, Jeffery, Ghazoul, Jaboury, Nelson, Paul, and Boedhihartono, Agni Kintuni. 2012. 'Oil palm expansion transforms tropical landscapes and livelihoods'. *Global Food Security* 1 no, 2: 114–119.

Schutz, Alfred and Luckmann, Thomas. 1973. *The Structures of the Life World*. Evanston, IL: Northwestern University Press.

Scott, James. 1985. *Weapons of the Weak*. New Haven, CT: Yale University Press.

Scott, James. 1998. *Seeing like a State*. New Haven, CT: Yale University Press.

Seabrook, Jeremy. 2014. *Pauperland: A Short History of Poverty in Great Britain*. London: Hurst.

Sillitoe, Paul and Marzano, Mariella. 2009. 'Future of indigenous knowledge research in development'. *Futures* 41: 13–23.

Smucker, Thomas and Wisner, Ben. 2008. 'Changing household responses to drought in Tharaka, Kenya: Vulnerability, persistence and challenge'. *Disasters* 32 no. 2: 190–215.

Smucker, Thomas and Wisner, Ben, et al. 2015. 'Differentiated Livelihoods, Local Institutions and the Adaptation Imperative: Assessing climate change adaptation policy in Tanzania. *Geoforum* 59: 39–50.

Stone, Moren and Nyaupane, Gyan. 2014. 'Rethinking community in community-based natural resource management'. *Community Development* 45 no. 1: 17–31.

Sultana, Farhana. 2011. 'Suffering for Water, Suffering from Water: Emotional geographies of resource access, control and conflict. *Geoforum* 42 no. 2: 163–172.

Survival International. 2016. 'Tribe attacked in palm oil plantation'. *Survival International*, 24 June https://www.survivalinternational.org/news/11340.

The Economist. 2015. 'What is austerity?' *The Economist*, Buttonwood's Notebook, 20 May. https://www.economist.com/buttonwoods-notebook/2015/05/20/what-is-austerity.

Trioni, M., Baciglupe, G., Gabrief, S. Knowles, G. Dickinson, S. Gil, M., Kelley, S., and Ludwig, J. 2019. 'Figuring Disasters, an experiment in thinking disruptions as methods' *Resilience* 7 no. 2: 192–211.

Tuan, Yi-Fu. 1979. *Landscapes of Fear*. Oxford: Blackwell.

Tuan, Yi-Fy. 1976. *Humanistic Geography*. Minneapolis: University of Minnesota.

Vickery, Jamie. 2018. 'Using an Intersectional Approach to Advance Understanding of Homeless Persons' Vulnerability to Disaster'. *Environmental Sociology* 4 no. 1: 136–147.

Waite, Louise. 2009. 'A Place and Space for a Critical Geography of Precarity'. *Geography Compass* 3, 1: 412–433.

White, Gilbert, Bradley, David, and White, Anne. 1972. *Drawers of Water: Domestic water use in East Africa*. Chicago: University of Chicago Press.

White, Gilbert. Bradley, David, and White, Anne. 2002. 'Assessing Domestic Water Use in Africa. Public Health Classics.' *Bulletin of the World Health Organization: The International Journal of Public Health* 80 no. 1: 63–73.

Wisner, Ben. 1978. *The Human Ecology of Drought in Eastern Kenya*. PhD thesis, Worcester, MA: Clark University.

Wisner, Ben. 1994. 'Jilaal, Gu, Hagaa, and Der: living with the Somali land, and living well'. In *The Somali Challenge: From Catastrophe to Renewal* edited by Ahmed Samatar, 27–64. Boulder: Lynne Reinner.

Wisner, Ben. 1995. '*Luta*, Livelihood, and Lifeworld in Contemporary Africa.' In *Ecological Resistance Movements*, edited by B. Taylor, 177–200. Albany: SUNY Press.
Wisner, Ben. 2010. 'Marginality'. In *Encyclopedia on Natural Hazards* edited by Peter Bobrowski, 651. Berlin: Springer.
Wisner, Ben. 2019a. 'R. W. Kates, human ecologist: the road more traveled and the road less traveled'. *Environment: Science and Policy for Sustainable Development* 61 no. 3: 39–44.
Wisner, Ben. 2019b. 'Speaking Truth to Power: A personal account of activist political ecology'. In *The Routledge Handbook of Political Ecology* edited by Thomas Perrault, Gavin Bridge, and James McCarthy, 53–63. London: Routledge.
Wisner, Ben, Blaikie, Piers, Cannon, Terry, and Davis, Ian. 2004 *At Risk*. 2nd Edition. London: Routledge.
Wisner, Ben, Gaillard, JC, and Kelman, Ilan. 2012. 'Framing Disaster: Theories and stories seeking to understand hazards, vulnerability and risk. In *The Routledge Handbook of Hazards and Disaster Risk Reduction* edited by Ben Wisner, JC Gaillard and Ilan Kelman, 18–34, London: Routledge.
Wisner, Ben, Heiman, Michael, and Weiner, Daniel 2005. 'Afterword: Jim Blaut, activist scholar'. *Antipode* 37 no, 5: 1045–1050.

11

DISASTER RISK CREATION

The new vulnerability

Thea Dickinson and Ian Burton

Humankind has long-held beliefs that ascribe disasters to *Mother Nature* or *Acts of God*. One variant asserted that the Acts of God were not fully independent of human actions but could be a form of retribution for unacceptable behaviour or sins. Noah's Flood is a prominent illustration. The recent conception of disaster risk creation echoes that biblical story by suggesting that human society can protect itself from disasters by changing and improving its behaviour and choices. The changes identified and needed now, however, are profoundly different.

Historical perspectives on vulnerability

The 1970s marked a pivotal period in the disaster vulnerability literature. A major new direction emerged; one that explored vulnerability as a result of human choices, decisions, and systems (Hewitt 1983; Hewitt 1997). This was in part built on the pioneering work of Gilbert White who advocated for a much wider range of choice in policies and methods to reduce flood damages. In 'Human Adjustment to Floods' (1945), White focused on a broad range of human choice rather than on the characteristics of the floods themselves, and the use of engineering technology to control them. This advance in understanding of disaster vulnerability laid the foundation for the growth of this interdisciplinary approach. Floods, White said, are 'Acts of God, but flood losses are largely Acts of Man' (Schwartz 2006; Cutter and Derakhshan 2019). While the work of White and his students (Burton 2001) led to the closer study of human choice as an explanation for disasters, it did not question or address the underlying root causes in the way that this is now being promoted (Burton 2010; Oliver-Smith et al. 2016).

In 1979, Westgate observed that 'any discussion of methods of mitigating future disaster losses sooner or later requires attention to be paid to land-use planning or management. Among long-term considerations, land-use planning

DOI: 10.4324/9781003219453-14

is often lauded as the most effective way of ensuring a less vulnerable future population' (Westgate 1979). This reasoning was an extension of White's flood management theory. In the section on Flood Fighting, while discussing levee construction, White mentions the relocation of the village of Espanola, New Mexico to a 'less vulnerable location'. He also comments on the flood vulnerability of Southern California (White 1945). While White and others focussed on the choice of adjustments to improve hazard management, they did not explore very deeply the processes and conditions that underlay the choices being made (Burton et al. 1978).

In 1992, Westcoat wrote, 'Hazards can be ameliorated through human adjustment. For White, the concept of adjustment encompasses adjustments of the environment, to the environment, and of human consciousness with response to the environment' (Wescoat 1992). The understanding of disaster vulnerability shifted from an external, uncontrollable entity to that which could be adjusted and responded to. 'Adjusting his use of a given part of a flood plain to conditions of flooding, man has the choice of occupying the land at the hazard of occasional flooding, of occupying it with partial or complete protection, and of staying out of the flood plain entirely' (White 1936).

However, by 1980 the use of the term vulnerability had garnered criticism. Vulnerability was labelled as an all-encompassing term with a lack of specificity and a great level of generality (Wisner and Luce 1993). The paper, 'Vulnerability, resilience and the collapse of society' (Timmerman 1981) was ahead of its time: The author comments, 'vulnerability is a term of such broad use as to be almost useless for careful description at the present, except as a rhetorical indicator of areas of greatest concern'. Despite this criticism, the use of the term 'vulnerability' as a concept in research and writing about disaster risks and losses continued to expand rapidly in the early 1990s (Wilches-Chaux 1992; Jeggle and Stephenson 1994; Davis 1994). By 1994 vulnerability had become a central tenet of research on natural hazards. A narrative of the emergence and growth of vulnerability is presented in the book 'At Risk' by Wisner, Blaikie, Cannon, and Davis (1994).

While there has been extensive research, literature, and practical application of the concept of disaster vulnerability, there has also been a rigidity within the field. Much of the work on vulnerability attempts to compartmentalise the facets. Contemporary narratives around disaster vulnerability are distracting attention from the growing threat of disaster risk creation (DRC). Economic and societal norms are generating disaster risk at a pace greater than society is implementing risk reduction. And given society's current realities – including the climate crisis and now pandemics – there is an even greater responsibility to expand our understanding of the complexities of disaster vulnerability.

Undoubtedly the work in the vulnerability domain has profoundly improved and strengthened the understanding of disasters. We know more, not just about the hazards or extreme events themselves, but also and especially, about the human choices involved in the creation of disasters (Oliver-Smith et al. 2016). We have also learned (and speculated) about the causes of vulnerability, but the knowledge gained has not been as widely or effectively applied as the scientific community might have expected (White et al. 2001).

There has been great success in the reduction of mortality and morbidity from disasters events up until 2020, while disaster losses continued to rise (Figure 11.1). Recognition that disasters not only continue to occur but are also increasing in magnitude and frequency has led to calls for a rethink and a re-examination of the concept of vulnerability (Bankoff 2019).

Figure 11.2 presents the global mortality data for disasters from 2000 to 2020. This graph omits losses of life attributed to SARS-CoV-2. Figure 11.3 is the data from Figure 11.2 with the addition of global mortality data for SARS-CoV-2.

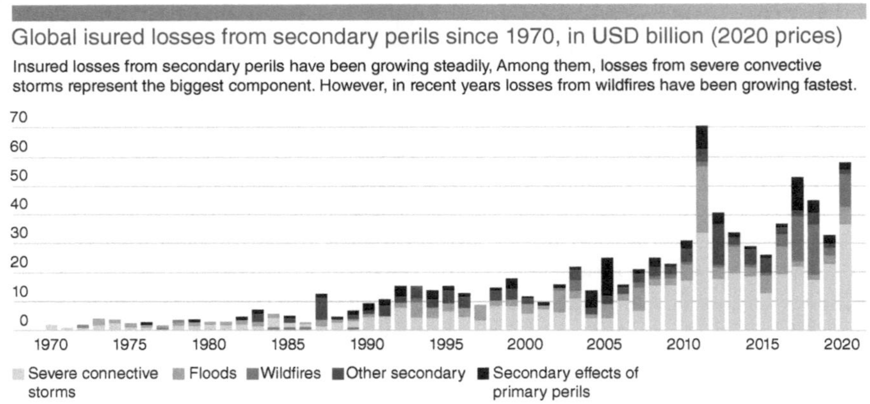

FIGURE 11.1 Global insured disaster losses from 1970 to 2020. Primary Perils are defined as tropical cyclones, earthquakes, winter storms; and Secondary Perils as river floods, torrential rainfall, landslides, thunderstorms, ice storms, drought, wildfires

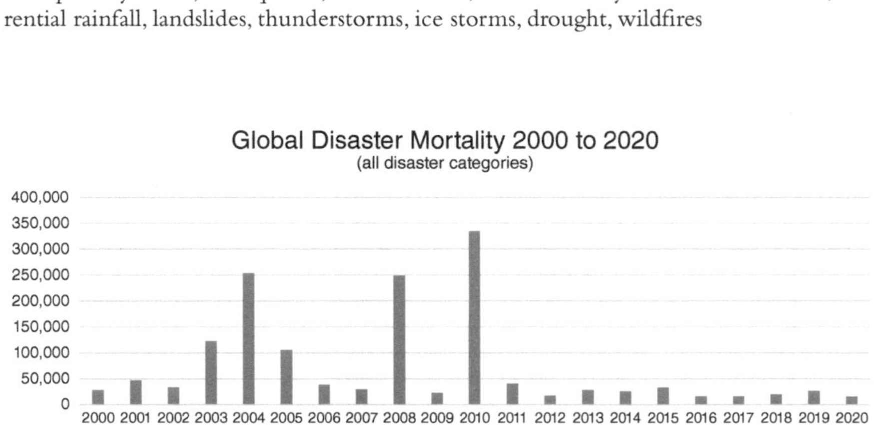

FIGURE 11.2 Global Mortality from Disasters 2000 to 2020. EM-DAT, CRED UCLouvain, Brussels, Belgium: www.emdat.be (D. Guha-Sapir) Accessed: February 2 2021

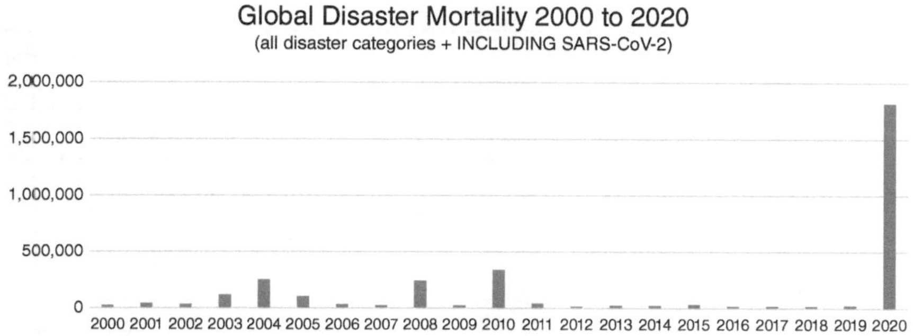

FIGURE 11.3 Global Mortality from Disasters 2000 to 2020, including losses attributed to SARS-CoV-2. SARS-CoV-2 data Max Roser, Hannah Ritchie, Esteban Ortiz-Ospina and Joe Hasell 2020: 'Coronavirus Pandemic (COVID-19)'. Published online at OurWorldInData.org. Retrieved from: https://ourworldindata.org/coronavirus Accessed: February 2 2021

The progress and advancement in vulnerability and disaster paradigms is presented in Table 11.1. While the table and the paper present the data as an evolutionary narrative, it does not always mean that the 'earlier' ideas or concepts have been replaced. In many cases, these differing narratives or dimensions of vulnerability and disasters still exist and co-exist in cultures, disciplines, media, and policy makers today. New ideas and perceptions tend to add to existing ones and not to replace them entirely. The phrases 'Acts of God' and 'natural disasters' persist in the media and much scientific, technical, and even legal literature. Considering this analysis, the term 'natural' applied to disasters is misleading and tends to focus public attention on the events itself (flood or earthquake, etc.) and the

TABLE 11.1 Vulnerability and disaster paradigms

Paradigm	*Acts of God*	*Disaster control and management*	*Disaster risk reduction*	*Disaster risk creation and future*
				Understanding
Language	Subject of natural forces	Natural Disasters	Natural and 'Man-made' disasters and hazards	Disasters (all types and subtypes, including epidemics and pandemics)
Cause	Overpowering forces of nature. Consequence of poor behaviours	Mother Nature and interaction.	External and Human Influence (interactive)	(mostly) Human: Actions, choices, and systems

(*Continued*)

TABLE 11.1 (Continued)

Paradigm	*Acts of God*	*Disaster control and management*	*Disaster risk reduction*	*Disaster risk creation and future*
				Understanding
Response	Accept dominant power of nature and/ or change behaviour	Disaster Management (technological monitoring and control of hazards)	Disaster Risk Reduction (vulnerability assessment) and Build Back Better.	Disaster Risk Creation (identifying/ reducing root causes of vulnerability and risk drivers; understanding, preventing, and creating accountability for DRC)
Scale	Place-based or local	Place-based or local	Community-based or regional	Global (with regional and localities)
	Single events in isolation	Single events in isolation	Disaster Subtype	Cascading, interconnected and common root causes.
	Sudden Onset Disasters	Sudden Onset Disasters	Sudden + Slow Onset (e.g., drought)	Sudden and Slow-onset disasters (inc. climate change and pandemics)
Accountability	Nature of God and human bad behaviour	Control over events, but partial and temporary.	Concern (Sendai) and	Responsibility (Disasters and Climate Change)
			Responsibility (UNFCC)	And human self-exposure to natural hazards (virus).
Epoch	Holocene	Holocene	Holocene/ Anthropocene	Anthropocene (natural hazards or extreme events have anthropogenic signal; natural variability exacerbated by Anthropocene)

emergency and humanitarian responses. It directs attention away from the root causes and responsibilities for the disaster losses.

The remainder of the chapter explores how our understanding and application of the knowledge of vulnerability has evolved and needs to continue to evolve and change in the face of the climate crisis, the COVID-19 pandemic, and increasing disaster losses.

From acts of God to anthropogenic root causes

When the term natural is used, it removes the element of human causation, shifting the responsibility away from human beings. In addition, responsibility of the powers that be (governments, corporations, society) is also removed by the language. This is not usually intentional, but often a side-effect of the decisions made by financial and governance systems in place. The result, however, has the potential to increase the level of disaster vulnerability, termed disaster risk creation (DRC). Lewis and Kelman (2012) define DRC as, 'processes that increase, or fail to decrease, vulnerability and that actively or inertially stymie or constrain [disaster risk reduction] efforts'. Until recently, DRC has gone unrecognised or unnoticed.

Legislation such as the US. 1936 Flood Control Act were important since they catalysed a shift from 'acts of god', or nature, to human control: the federal government was assuming responsibility for flood protection (while Gilbert White would note, there were issues with this: the 'man over nature' paradigm, rather than an interactive or synergistic relationship). By the 1980s the view of the attribution of disasters started shifting. In 1983, Hewitt stated, 'the dominant view, then, disaster itself is attributed to nature. There is, however, an equally strong conviction that something can be done about disaster by society' (Hewitt 1983). By this time, 'much of the literature use[d] the terms 'disaster' and 'natural disaster' synonymously (Berren et al. 1980). And a new field, human ecology of hazards, emerged attempting to 'explain the differential effects of hazards not only on physical structures but on people, their economic activities and social relationships' (Maskrey 1989).

The contemporary narrative rests somewhere between an interactive causation of external factors and the idea of total human influence. From the early 1970s on, the weight accorded to human actions and choices has been steadily increasing to the point where it has sometimes been suggested that disaster losses from extreme natural events are entirely attributable to human choice (Oliver-Smith et al. 2016). The future of disaster vulnerability must continue to evolve to understand how human 'actions', 'choices', and 'systems' are creating increased levels of disaster risk.

A parallel shift was occurring within the disaster risk field. Inherently, natural disasters only included hazards categorised as meteorological, geophysical, or hydrological. When the United Nations Office for Disaster Risk Reduction

(UNDRR) set out to develop the Sendai Framework the risk models in common usage were revisited that underpinned their directive. A 'conceptual shift' was required as the current list of hazards and risk included under the mandate 'did not reflect the full range of risks'. Consequently, the qualification of 'natural' was removed and technological and biological hazards were adopted (GRAF 2019).[1]

Timmerman's 1981 observation of vulnerability was that it had become a term of such broad use that it had less, and less degrees of accuracy or precision. There is also, however, concern that vulnerability research takes place in disciplinary silos and neglects cross-scale interactions. Belief systems, language use, and attributed causation perpetuate these silos. Disasters are complex, cascading, and interconnected by underlying causes, and therefore a broadening of comprehension is required, but not one that simply creates generality, but rather increases our understanding.

Accountability: common concern and responsibility

The historical (and still persistent) narrative that disasters are natural and external from human causes has allowed for a level of responsibility to be evaded, both within international policy frameworks, as well as nationally and locally. The progress in dealing with anthropogenic climate change provides a revealing and instructive comparison. From the outset climate change was recognised as a global problem. All nations and all people contribute to varying degrees to the emission of greenhouse gasses, and all people are at some differential risk from the impacts of climate change. The level of contributions to the ongoing creation of the problem and exposure to risk are not equal, but they are recognised as universal. The atmosphere is everywhere and belongs to all. It is a global common property resource and 'responsibility'.

Globalisation has heightened the ability for local, and perhaps historically previously contained disasters, to cascade into other localities and, as demonstrated by the COVID-19 pandemic, to spread globally. The understanding of this interconnection is necessary to the understanding and reduction of disaster risk creation. The question remains, however, who bears responsibility, and where the level of common concern lies. The global commons are understood as a universal responsibility; the same has yet to be defined for disasters with local origin that spread wildly beyond their borders. The intersection of localities, national authorities, and international organisations blurs responsibility. A key to reducing disaster risk creation is a globally recognised level of common concern, accountability, responsibility.

This acceptance of the idea of responsibility was written into the text of the UN Framework Convention for Climate Change (UNFCCC) opened for signature in Rio de Janeiro in 1992 at the World Conference on Environment and Development. The Preamble to the Convention includes the following statement:

> The Parties should protect the climate system for the benefit of present and future generations of humankind, on the basis of equity and in accordance with their common but differentiated responsibilities and respective capabilities.
> (UNFCCC Article 3 Principles, Clause 1)

The Convention then goes on to list several objectives, principles, and commitments. The phrase 'common but differentiated responsibilities' is key to the negotiations and agreements (including the Paris Agreement of December 2015), that have been reached since the Framework Convention came into force in (1995). At the time of writing (in early 2021) climate change is recognised among all national governments including the USA as a 'common responsibility'. By contrast, in the negotiations and draft texts of the international frameworks on disasters the word 'responsibility' has been applied only to nation states within their own jurisdiction and not to the collective body of all nations or the international community. Opposition to the recognition of disasters as a common global responsibility comes overwhelmingly from the richest and more highly developed western countries. The phrase 'common concern' has usually been used as a replacement. Section III of the Sendai Framework states in paragraph 19 a: (a)

> Each State has the primary responsibility to prevent and reduce disaster risk, including through international, regional, subregional, transboundary, and bilateral cooperation. The reduction of disaster risk is a common concern for all States and the extent to which developing countries are able to effectively enhance and implement national disaster risk reduction policies and measures in the context of their respective circumstances and capabilities can be further enhanced through the provision of sustainable international cooperation.

Note that 'sustainable international cooperation' does not specify any financial assistance or cost-sharing.

The question then comes about, why climate change is a common 'responsibility' and the disaster problem only a common 'concern'? If climate change is a responsibility, but also categorised as a disaster (e.g., extreme weather events attributed to or enhanced in terms of magnitude and probability by climate change), then logically disasters too must be a common responsibility. However, it is argued that climate change is a global problem to which all nations and peoples contribute, and to which all are exposed, whereas disasters are place-based in time and space. The previous sections demonstrate that this is a historical narrative that is being continued. As knowledge and understanding increases about the root causes of disasters it becomes exceedingly difficult to perpetuate the 'common concern' view. As the connectedness of disasters is increasingly recognised, especially in the underlying or root causes that create unsafe conditions and dynamic pressures, then the case for recognising disasters as a common responsibility becomes steadily stronger.

Vulnerability is not created by a clearly identified pollutant like carbon dioxide and other greenhouse gasses, but it is being created through pervasive and global development processes. Furthermore, extreme atmosphere-related events are increasingly recognised by the scientific community (Intergovernmental Panel on Climate Change, IPCC) as increasing in magnitude and frequency partly due to climate change (Field et al. 2012). One large obstacle is undoubtedly the recognition that with responsibility comes cost. The UNFCCC (Article 4.4) contains the commitment that:

> The developed country Parties and other developed Parties included in Annex II shall also assist the developing country Parties that are particularly vulnerable to the adverse effects of climate change in meeting costs of adaptation to those adverse effects.
>
> (UNFCCC Article 4 Commitments, Clause 4)

Under this commitment the developed countries have made substantial promises of funds, and some funds have been made available and used. The Sendai Framework for Disaster Risk Reduction 2015–2030 is the successor instrument to the Hyogo Framework for Action (HFA) 2005–2015: Building the Resilience of Nations and Communities to Disasters. The HFA was conceived to give further impetus to the global work under several agreements: the International Framework for Action for the International Decade for Natural Disaster Reduction of 1989, (the Decade was from 1990 to1999); the Yokohama Strategy for a Safer World: Guidelines for Natural Disaster Prevention, Preparedness and Mitigation and its Plan of Action, adopted in 1994; the International Strategy for Disaster Reduction of 1999.

As the conferences and agreements and frameworks continue, there are changes in language and terminology. For example, the idea of 'natural disasters' has now been rejected in favour of the recognition that disasters result from human choices. Recall that Gilbert White wrote in 1945 that 'flood losses are largely acts of man'. National delegations meet and agree that progress has been made, and then formulate another framework. The dominant assumption is that disasters are to be managed and the risks reduced primarily by the nations and communities where they occur. The role of the global international community is to be advisory, prescriptive, supportive, and to enhance cooperation, recognising that developing countries 'require an enhanced provision of means of implementation, including adequate, sustainable and timely resources, through international cooperation and in global partnerships for development and continued international support so as to strengthen their efforts to reduce disaster risk'.

Humanitarian aid and contributions to the costs of rehabilitation and reconstruction are still by far the dominant components of international assistance. Such assistance is voluntary and 'humanitarian'. There is no sense of legal obligation or responsibility. In some cases, no doubt, there is a sense of moral responsibility.

In the end disaster assistance is a charitable act, not an obligation. It results from concern and not responsibility.

This is far from the UNFCCC and the Paris Agreement of 2015. There is no recognition of a global level of responsibility other than that nations and the international community and organisations should do something, preferably together. But then the UNFCCC is a Convention whereas Sendai and its predecessors are just 'frameworks'. It is all rather wishful thinking, but at least a Convention has a higher status in international 'soft law' and is voluntarily 'binding'.

The struggle to create genuine and legally 'binding' agreements and governance at a global level for disasters goes on. Undoubtedly the resistance to financial commitments from the donor countries has been the main reason for insisting on 'common concern' rather than 'common responsibility'. In other ways the rhetoric and the negotiated text tries to keep pace with the unfolding and deteriorating situation. The action falls well short.

Vulnerability has come to occupy a central place in the climate change debates. Not surprisingly the use of the term has followed a similar trajectory to that in the field of disasters. There is also a curious circularity in this evolution. To ascribe losses and disasters to acts of god which in their turn have been triggered by human misbehaviour, sounds remarkably like the current argument that human actions and choices are to blame. Climate change currently tops the list of disastrous threats, and this is attributed to human actions and choices. The greatest threat to humanity is us. As the cartoon character Pogo presciently said many years ago, 'We have met the enemy, and he is us' (Kelly 1972).

Disaster risk creation

The latest and current global-level disaster is the COVID-19 (and variants) pandemic. Research and investigations into the cause and potential responsibility for the pandemic are still ongoing (early 2021) and a clear and definitive answer may never be achieved. It seems clear however that the virus originated in an animal species. Although this is being questioned by some scientists and epidemiologists, bats via an intermediary animal are the prime suspect. Like animal-to-human transmission of Ebola and HIV via bushmeat, so-called 'wet markets' where live animals are on sale for use as human foods and traditional medicines are the presumed location of first transmission to humans.

The identification of wet markets as a place of transmission has led to calls for their abolition. Considering their widespread existence in parts of the global south, and also in some more 'developed' countries, such an action would be difficult to achieve in the short term, and in any case would not address the many potential modes of transmission and spread. Reasons for the rapid global spread of the COVID 19 pandemic include the lack of preparedness, and the often slow recognition of the degree of danger. Even once the pandemic's rapid spread was recognised, the response was poorly managed in many countries (but not all), and the processes of testing and vaccination, lockdown, and social distancing, have been disorganised and confusing to say the least. Some of this can be attributed to scientific uncertainty about the nature of the virus, and the changing scientific advice about what actions to take and when to relax them.

It is widely agreed that scale and rapid spread of the COVID 19 pandemic represents a remarkable failure of human response considering the number of epidemics and potential pandemics that have occurred in recent years. This includes the ongoing HIV – AIDS and the 2003 Severe Acute Respiratory Syndrome (SARS). Despite these experiences, preparations to deal with pandemics have been relaxed and reduced. As in the case of other disasters both the initiating event and the magnitude of the consequences are both attributable largely or perhaps entirely to human choices, action, and inaction.

The future of vulnerability

The shift from the geological epoch of the Holocene to what is currently being termed the Anthropocene marks a stage in planetary evolution in which humanity has become the dominant force. Among the consequences are the human actions, choices, and systems that generate disasters at a pace faster than the measures being put into place to decrease risk. The future of vulnerability is dependent on human capacity, intent, and will to act upon three interconnected elements of the disaster problem. The top priorities of concern (not yet all a recognised collective global responsibility) are climate change, pandemics, and globalisation. These three catalysts are forcing the world to change the understanding of disaster vulnerability, while at the same time continuing to cause increases in disaster vulnerability. Climate change and pandemics and the underlying processes of development are not independent. There are feedback loops opening the way to catastrophe and existential threats. The human world and the planet are in a precarious situation.

Disaster vulnerability cannot be viewed through the same lens as it was in the 1970s. The meaning of the concept of 'vulnerability' as applied to extreme events and disasters has radically changed over the last four decades. Initially, it referred to small groups and individuals who were deemed to be vulnerable due to their own personal circumstances such as poverty, health, age, gender, mobility, quality of housing, and location. In this use of the term, vulnerability was seen as being restricted in both time and space. Through this lens, vulnerability could therefore be reduced by keeping infrastructure, housing, and public facilities such as hospitals and schools out of places or areas subject to the hazards of extreme events. It could also be reduced by addressing the personal circumstances of those vulnerable or at risk, including poverty, health, and so forth.

This thinking has underlain much of the recent effort both at national and global levels embraced by the idea and the goal of 'disaster risk reduction'. While the current efforts directed at disaster risk reduction are valuable and worthwhile, they fail to capture the full dimensions of the problem of disaster risk creation.

Risk creation is a key element missing from the disaster-vulnerability paradigm. DRC is a process that increases vulnerability. The future of vulnerability is the observation and recognition of the key role DRC plays in creating unnecessary increases in vulnerability and disasters. Other dimensions which are now becoming increasingly apparent include the cascading of disaster impacts and a slowly dawning recognition that disasters cannot be seen only as isolated place-based events but also result from common underlying processes (Dickinson and Burton 2015). Vulnerability now must include both the world and the planet. The world is in a precarious state and the term 'precarity' is being added to the vocabulary of vulnerability experts (see Hewitt paper in this book).

Conclusion

Of growing salience among a range of underlying factors and root causes are climate change, pandemics, and economic and technical globalisation. Long gone are the days when it seemed reasonable to refer to 'natural' disasters because the disasters in question were caused by extreme natural events. It has come to be understood, after many decades of debate, that disasters are the result of human choices which help to create and perpetuate local and personal vulnerability. With the advent of climate change, and the role of climate change in destabilising atmospheric processes, and increasing the frequency and magnitude of extreme events, and the emergence of epidemics and pandemics, it is now apparent that underlying causes of disasters include the whole global system of economic growth and development. Vulnerability is no longer only, or mainly, localised, it is widespread and global.

The world population now lives in an era of global vulnerability and the concept of disaster risk reduction must be broadened to include the actions and processes which are creating the conditions for disaster. These conditions are being created at a faster rate than the present efforts at disaster risk reduction. Vulnerability at a global level requires re-examinations of the locations of responsibility for policy and action. It is no longer sufficient to rest the responsibility for the prevention and management of disasters at the local and national levels and limiting international action to humanitarian responses of forecasts, warnings, evacuation, relief, and reconstruction. There needs to be recognition of a new collective and global responsibility for 'natural' disasters in the way that this now exists for climate change. The use of the concept of vulnerability must be expanded to bring about new and more effective responses to the processes creating the new reality of global vulnerability.

Note

1 Within the Sendai Framework the list of risk and hazards now includes the following: Natural hazards, Environmental risks, Technological risks, Biological risks, Human-induced risks.

References

Bankoff, Greg. 2019. Remaking the world in our own image: vulnerability, resilience and adaptation as historical discourses. *Disasters*, 43 (2), 221–239.

Berren, Michael R., Allan Beigel, and Stuart Ghertner. 1980. A typology for the classification of disasters. *Community Mental Health Journal*, 16(2), 103–111.

Burton, Ian. 2001. The intergovernmental panel on natural disasters. *Environmental Hazards*, 3, 139–141.

Burton, Ian. 2010. Forensic disaster investigations in depth: A new case study model. *Environment*, 52, 5, pp. 36–41.

Burton, Ian, Kates, Robert, and Gilbert White. 1978. *The Environment as Hazard*. Oxford: Oxford University Press.

Cutter, Susan L., and Sahar Derakhshan. 2019. Implementing disaster policy: Exploring scale and measurement schemes for disaster resilience. *Journal of Homeland Security and Emergency Management*, 16(3). doi: 10.1515/jhsem-2018-0029.

Davis, Ian. 1994. Assessing community vulnerability. In *UK National Coordination Committee for the International Decade for Natural Disaster Reduction. (IDNDR)* (ed.) Medicine in the International Decade for Natural Disaster Reduction (IDNDR) pp. 11–13. Proceedings of a Workshop at the Royal Society, London. 19 April 1993.

Dickinson, Thea, and Ian Burton. 2015. The disaster epidemic: Research, diagnosis, and prescriptions. In *Risk Governance* edited by Paleo U. Fra, 185–200. Dordrecht: Springer.

Field, Christopher B., Vicente Barros, Thomas F. Stocker, and Qin Dahe. eds. 2012. *Managing the Risks of Extreme Events and Disasters to Advance Climate Change Adaptation: Special Report of the Intergovernmental Panel on Climate Change*. Cambridge, UK: Cambridge University Press.

GRAF. 2019. *The Global Risk Assessment Framework: Better Decisions for a Better Future*, Concept Note. https://www.preventionweb.net/files/61909_grafnewdraftconcept-note.pdf.

Hewitt, Kenneth. 1983. *Interpretations of Calamity*. Boston. Allen and Unwin.

Hewitt, Kenneth. 1997. *Regions of Risk: A Geographical Introduction to Disasters*. Harlow, UK. Longman.

Jeggle, Terry, and Stephenson, R. 1994. Concepts of hazard and vulnerability analysis. In *Disaster Management* edited by V. Sharma, 251–257. New Delhi: Indian Institute of Public Administration.

Kelly, Walt. 1972. Pogo: We Have Met the Enemy and He is Us. New York: Simon and Schuster.

Lewis, James, and Kelman, Ilan. 2012. The good, the bad and the ugly: Disaster risk reduction (DRR) versus disaster risk creation (DRC). *PLOS Currents Disasters*, 4. doi: 10.1371/4f8d4eaec6af8.

Maskrey, Andrew. 1989. *Disaster Mitigation: A Community-Based Approach*. Nairobi, Kenya: Oxfam International.

Oliver-Smith, Anthony, Irasema Alcantara-Ayala, Ian Burton, and Allen Lavell. 2016. *Forensic Investigations of Disasters (FORIN): A Conceptual Framework and Guide to Research*. Beijing: Integrated Research on Disaster Risk. ICSU, 56 p.

Roser, M., Ritchie, H., Ortiz-Ospina, E., and Hasell, J. 2020. Coronavirus pandemic (COVID-19). *Our world in data*.

Schwartz, John. 2006. Gilbert F. White, Obituary. *New York Times*. October 7.

Timmerman, Peter. 1981. *Vulnerability, Resilience and the Collapse of Society: A Review of Models and Possible Climatic Applications* (No. 1). Toronto, Canada: Institute for Environmental Studies, University of Toronto.

Wescoat, James. 1992. Common themes in the work of Gilbert White and John Dewey: A pragmatic appraisal. *Annals of the Association of American Geographers*, 82(4), 587–607.

Westgate, Ken. 1979. Land-use planning, vulnerability and the low-income dwelling. *Disasters*, 3(3), 244–248.

White, Gilbert. 1936. The limit of economic justification for flood protection. *The Journal of Land and Public Utility Economics*, 12(2), 133–148.

White, Gilbert. 1945. *Human Adjustment to Floods*. Chicago: University of Chicago, Department of Geography Research Paper, No. 29.

White, Gilbert. 1958. *Changes in Urban Occupance of Flood Plains in the United States* (Vol. 57). Chicago: University of Chicago.

White, Gilbert, Robert Kates, and Ian Burton. 2001. Knowing better and losing even more: The use of knowledge in hazards management. *Global Environmental Change Part B: Environmental Hazards*, 3(3), 81–92.

Wilches-Chaux, G. 1992. La Vulnerabilidad. En Maskrey, *A. Los desastres no son naturales.*

Wilches-Chaux, G. 1993. La Vulnerabilidad Global. *In Los desastres no son naturales*, edited by Andrew Maskrey. Bogotá. La Red, Lima. https://www.desenredando.org/public/libros/1993/ldnsn/LosDesastresNoSonNaturales-1.0.0.pdf.

Wisner, Ben, and Henry R. Luce. 1993. Disaster vulnerability: Scale, power and daily life. *GeoJournal*, *30*(2), 127–140.

12

VULNERABLE ANTHROPOCENES?

Towards an integrated approach

Kasia Mika and Ilan Kelman

The 'Anthropocene' and beyond

The scene of '-cenes'

As discussed by scholars from across social sciences and the humanities, the 'Anthropocene', from its public journal appearance onwards (Crutzen and Stoermer 2000, 2002) evokes an increasing number of divergent, if not contrasting, ideas and approaches to issues of environmental change, human agency, geologic times and human and non-human futures. Whether seen as a boundary (Haraway 2015) – 'more a boundary event than an epoch, like the K-Pg boundary between the Cretaceous and the Paleogene' (Haraway 2015, 160) – or an epoch best demarcated through stratigraphic evidence and marking points (Davies 2016), the Anthropocene functions in different ways across its many uses and definitions. It also becomes a cultural notion capturing the rate of anthropogenic environmental change, a political call to action or, for some, a relevant, non-anthropocentric category which begins 'when the characteristic conditions of the Holocene no longer exist' (Davies 2016, 51).

For all its growing fame, and the concomitant, wide-varying critique (e.g., Chakrabarty 2009; Malm and Hornborg 2014), the Anthropocene is not the only boundary or epoch demarcating notion which has been proposed. Moving away from debates on chronology, concerned at the most basic level with establishing an ever-clearer definition and starting point for the claimed-to-be new epoch, the proliferation of alternative or supplemental terms seems to suggest the insufficiency of the 'anthropos' as the root (Usher 2016, 57–58) of a new geological and cultural era. Plantationocene (Haraway et al. 2015), Capitalocene (Malm 2016; Haraway 2016; Moore 2017, 2018), and Chthulucene (Haraway 2015) are among the most prominent. Others include Agnotocene, Anthrobscene,

DOI: 10.4324/9781003219453-15

Eremocene, Homogenocene, Manthropocene, Naufragocene, Northropocene, Outiscene, Phagocene, Polemocene, Phronocene, Thalassocene, Thanatocene, and Thermocene (Usher 2016, 58; n5, 70) as well as Plasticene in English (Reed 2015) and Poubellocène in French (Monsaingeon 2017) from *poubelle* in French meaning waste and trash. These are some, but not all, of the terms proposed to capture and mobilise a critical rethinking of the sense of a joint, societal/cultural, ecological/environmental, and political/governance crisis in which we find ourselves, alongside the consecutive demand to reconfigure our place in the world and the range of human and non-human relations in which we are entangled. As Ginn et al. (2018, 214) succinctly put it:

> For all their differences, the new –cenes of the environmental humanities and the critical social scientists seem united in arguing that science alone cannot reveal the fractured timespace of our present planetary moment.

While capturing the fundamental methodological tensions between and within the divergent calls for renaming and inaugurating a new epoch, Ginn et al. (2018, 214) effectively underline the key stakes behind these calls, namely: an integrated and longitudinal approach to 'the fractured timescape of our present planetary moment.' Similarly, in an earlier conversation on the different '-cenes' and their respective emphases – whether it is an emphasis on capital, making kin with non-human others, or colonialism and the plantation economy – Haraway (2015, 27) cautioned against these '-cene' theories becoming, too big, too fast, and falsely implying 'a theory of everything'. Elizabeth DeLoughrey (2019) also warns against the universalising discourses discernible in Anthropocene scholarship and the seemingly universal, spatio-temporal scope of the notion. Echoing Dipesh Chakrabarty's earlier call to 'provincialize Europe' (2000), she argues 'that must "provincialize" the Anthropocene [...since] the "universal and secular vision of the human" that Chakrabarty sought to decenter in European discourses has been regenerated in much Anthropocene scholarship of the "Age of Man," resurrecting a figure who reigns as a singular (masculine) 'species' (DeLoughrey 2019, 2–3). While acknowledging the urgency and complexity of the current challenges signposted across the different terms, avoiding the 'Buzzwordocene' through a focus on vulnerability, as this chapter discusses, allows for a methodologically and analytically integral and critical take, one that encompasses multiple times, spaces, and factors.

Vulnerability-cene: towards an integrated view

This integrated, vulnerability-focused view, which resists the privileging of one main historical marker or characteristic (e.g., the plantation economy or the scale of plastic waste), is formulated in terms of methodology and analytical scope. It recognises the necessity for a truly multidisciplinary take by encompassing diverse knowledge forms, histories, and cultures. Vernacular, traditional, local, and indigenous approaches are embraced, also encompassing the expertise of scientists, professionals, and artists. Modern, scientific knowledge includes physical sciences, social sciences, and the humanities. Professionals encompass engineering, social work, law, architecture, and medicine as well as those working in creative and performing arts. All these intersect and interact, with varying levels of disciplinary divisions and interdisciplinary collaborations. A vulnerability-focused approach values these varying knowledge cultures equitably and seeks to join their respective insights in order to understand and address contemporary challenges while still allowing for a critical analysis of 'forms of knowledge attendant on colonial modernity' (Shepherd 2018; see also Mignolo 2008, 2011, 2013). It does so by bringing together different fields, one example of which is climate change and disaster risk reduction, and actors, from researchers to government bodies and local communities, navigating the respective challenges from differing standpoints, levels and forms of engagement.

Simultaneously, it is important to note that the chapter's emphasis on vulnerability, and a move away from the debates on the scope, relevance, and limitations of consecutive '-cenes', is by no means a move away from the concerns of ecological, environment, and social justice (embracing climate justice and disaster justice), or the notions of agency, responsibility, equity, and planetary futures that are at their core. Neither is this an escapist's or a denier's call to just 'keep calm and carry on'. There are real, highly damaging, and deeply embedded ecological, social, cultural, historical and political problems, clearly signposted in the prefixes to the different '-cenes', which need to be addressed and which are obviously not limited to the neat definitions of 'cut-off points'. These are as basic as a perverse economic system which often subsidises fossil fuels while punishing renewables; a taxation system which tends to favour those who are already rich; and inherent inequities which typically stop many people from achieving their fullest contributions to society; or the systemic legacy of colonialism and imperialism which is signalled in many of these '-cenes' and which continues to shape communities' lives and futures.

In this context of overlapping and longitudinal factors, a re-anchoring of discussions of climate change within the notion of vulnerability allows us attend to, in a more rigorous and politicised manner, these very questions of justice, responsibility, agency, and futures. It does so by bringing together, rather than splitting into ever smaller camps, the parallel critical efforts signposted by each of the proposed '-cenes' as well as the work undertaken, too often seen as disconnected rather than overlapping, in other disciplines under topics such as 'disaster risk reduction' and 'climate change'. In other words, this integrated approach that the chapter proposes turns to the wider idiom of vulnerability and justice, in order to reconnect the oft parallel, and, at times, seen as competing, critical efforts of scholars from across environmental humanities, postcolonial studies, political ecology and those working in disaster risk reduction and climate change. In effect, by drawing out the common concerns from across humanities and (social) sciences, the chapter weaves together the different scholarly languages, including their respective vocabularies on climate change and disaster risk reduction, in an attempt to re-anchor the discussions of lives and futures in a political commitment to justice and vulnerability reduction.

Overview of climate change

To achieve this chapter's goal, a baseline understanding of climate change and of the term's dominant uses is needed. The UN provides two different definitions for contemporary climate change. The UN has one body called the Intergovernmental Panel on Climate Change (IPCC) which is responsible for synthesising and assessing, in a balanced manner, the current understanding of the science of climate change. Their reports then go through review and approval by member state governments. The first assessment was in 1990 and the most recent published one was the fifth assessment from 2013 to 2014 with the sixth assessment currently underway. The definition of 'climate change' in IPCC (2013–2014, Glossary, 5) is:

> Climate change refers to a change in the state of the climate that can be identified (e.g., by using statistical tests) by changes in the mean and/or the variability of its properties, and that persists for an extended period, typically decades or longer. Climate change may be due to natural internal processes or external forcings such as modulations of the solar cycles, volcanic eruptions, and persistent anthropogenic changes in the composition of the atmosphere or in land use.

In parallel, the main international agreement for addressing climate change is the United Nations Framework Convention on Climate Change (UNFCCC), the secretariat of which now uses the acronym UNFCCC to refer to itself as UN Climate Change. UNFCCC, 3, Article 1, Paragraph 2) defines climate change as 'change of climate which is attributed directly or indirectly to human activity that alters the composition of the global atmosphere and which is in addition to natural climate variability observed over comparable time periods.' The definitional difference emerges in that the government-reviewed scientific synthesis and assessment through the IPCC (2013–2014) considers all changes to the climate, no matter why those changes are occurring, whereas the international policy approach (UNFCCC 1992) narrows its mandate to only climate change caused by human activity. Nonetheless, both bodies accept that anthropogenic climate change is projected to lead to a global climate which is vastly different from what humanity has experienced before.

Yet climate itself is not and cannot typically be a disaster per se, because a disaster by definition cannot happen without aspects or elements of society being affected (Hewitt 1983, 1997; Lewis 1999; UNISDR 2009, 2017; Wisner et al. 2004). Climate, by definition, is average weather based on statistical analyses, so it includes potential environmental hazards such as floods, storms, temperature extremes, wildfires, and droughts. None of these phenomena represents an unusual environmental process. For example, floods, wildfires, and droughts are part of ecosystems with many species adapted to their regular occurrence. In fact, they are not even necessarily hazardous unless unprepared people and our infrastructure get in the way, representing the importance of vulnerability.

People can choose to live in harm's way, for instance by purchasing a house in the floodplain to enjoy beautiful views of the river without taking adequate measures to reduce flood damage. More often, people are forced into vulnerable situations such as choosing not to evacuate because of the fear and threat of sexual violence in the public storm shelter. That is, vulnerability is a social and political process of human decisions and choices making disasters a social and political process encompassing human decisions, choices, and omissions (Cannon 1994; Gaillard 2019; Hewitt 1983, 1997; Krüger et al. 2015; Lewis 1979, 1999; Wisner 1993; Wisner et al. 2004). Climate change – by IPCC and UNFCCC definitions – can most significantly influence hazards rather than vulnerabilities. Climate change will make some hazards worse, will reduce potential problems from other hazards, and will have no influence on further sets of hazards. One example is that many storms including tropical cyclones seem to be decreasing in frequency but increasing in intensity due to climate change (e.g., Romero and Emanuel 2017; Knutson et al. 2020).

The separation between hazards and vulnerabilities, though, is not always clear-cut, and there is sometimes a considerable overlap between them. Climate change will also push some hazards into realms that preclude or severely curtail human survival in large swathes around the planet. The most prominent is heat waves. The combination of heat, humidity, and time above certain values – which vary according to demographic characteristics and any measures taken to prevent adverse health impacts – leads to most mortality in heat waves, because it becomes dangerous to health to be outdoors or to experience the heat-humidity combination for several days in a row, even with continual hydration (Watts et al. 2020). Climate change is pushing heat waves into this realm and, unless climate change is tackled, these situations will become increasingly common, also substantially inhibiting agriculture, food systems, and freshwater supplies. Another example would be ice sheet collapse leading to dozens of metres of sea-level rise (Clark et al. 2016), meaning that people would either have to move from inundated land or reconstruct settlements in the inundated locations.

Yet climate change is only one aspect of the changing environment from human and non-human influences – and it is not clear that climate change is necessarily the largest change or the change that is most important for overall vulnerability levels. The literature is rife with other examples of major human-induced global changes such as omnipresent microplastics, loss of fertile topsoil, freshwater depletion, chemical load, and species and ecosystem losses. Together, these rapid, wide-scale alterations are exposing the long-standing, ever-present, chronic vulnerability conditions which form the baseline of the social ills experienced, such as: inequity, injustice, discrimination, and oppression. Climate change is one more exposer and factor among the many complex, historical as well as recent, processes underlying vulnerability. This narrowing of the application and scope of the term, as one among many factors, might seem to be in stark contrast to the humanities' much more expansive use of 'climate change.' DeLoughrey's own definition encapsulates this methodological divergence: 'my definition of "climate change,"' she writes, 'refers to a world-changing rupture in a social and ecological system that might be read as colonization in one context or sea-level rise in another' (DeLoughrey 2019, 7). Whereas such spatio-temporal extension and substitution might at first seem impossible to reconcile for some scholars in (social) sciences, what both approaches capture, each drawing on their disciplinary language, is the necessity to rigorously pluralise and recalibrate the factors and the types of analyses brought to bear on climate change, its history and present manifestations, vulnerability, and disaster risk reduction.

This ethos of 'multiple exposure' and overlapping factors founded many aspects of disasters and development work, especially placing disasters within development contexts (Cannon 1994; Glantz 1976; Hewitt 1983, 1997; Krüger et al. 2015; Lewis 1979, 1999; O'Keefe et al. 1976; Torry 1979; Wisner 1993; Wisner et al. 2004). Contemporary anthropogenic climate change, in its first narrower UN use, adds a layer to these discussions, but does not change the fundamental, broad-based concerns of what vulnerability is, how it arises, and why it is not being addressed – concerns which were discussed in the literature long before climate change reached the international policy agenda through the IPCC being founded in 1988 and the UNFCCC being founded in 1992. The key from such work is that those who are most vulnerable to, and least able to cope with, one challenge, tend to be most vulnerable to, and least able to cope with, multiple challenges. They are exposed to multiple challenges simultaneously which, in turn, significantly inhibits them from improving their situation without external assistance. Climate change as one influencer of hazards and the environment amongst many (especially given its baseline definition of changes in weather statistics) therefore should not be separated from other topics (Kelman 2017). It becomes part of the wider concerns articulated about equality, equity, the environment, and human-environment interactions. These concerns, to some degree, are encapsulated by the concept of vulnerability.

Vulnerability

Integrated efforts

Vulnerability brings back together many issues which have been separated or siloised, joining the various emphases of the respective 'cenes' in order to produce and lead joint, rather than parallel, efforts. Examples of the parallel efforts are having separate processes, procedures, outcomes, and agreements for:

- Climate change: UNFCCC meetings led to the Kyoto Protocol (UNFCCC 1997) and the Paris Agreement (UNFCCC 2015).

- Disaster risk reduction: The Hyogo Framework for Action (UNISDR 2005) was followed by the Sendai Framework for Disaster Risk Reduction (UNISDR 2015).
- Sustainable development: The Millennium Development Goals (UN 2000) were succeeded by the Sustainable Development Goals (UNGA 2015).

All three, and others in parallel, also had long histories prior to the documents cited in the bullet points. Indeed, some efforts are made at integration, demonstrating positive successes. For instance, disaster-related work does not have its own Sustainable Development Goal with disaster efforts instead distributed amongst relevant goals. In contrast, climate change has its own goal which is explicitly stated as being within UNFCCC's mandate, thereby separating climate change and sustainable development, even while some other sustainable development goals impressively integrate action related to climate change.

Vulnerability becomes an integrator in this regard because it seeks baseline understandings and explainers, capturing a wider variety of influences and impacts than any single topic can achieve. A vulnerability focus allows identifying fundamental causes and tackling them more effectively, and thereby alleviating numerous symptoms, such as disaster risk, deforestation, overfishing, and anthropogenic climate change. Moreover, in terms of a conceptual interdisciplinary engagement with complex and overlapping issues, vulnerability offers a more rigorous and open-ended frame than that offered by analyses separating these interconnected topics or those separating climate change from wider socio-political processes and environmental frames. The open-endedness refers to recognising multiple spatial, temporal, and governance scales, including but not limited to the various conceptualisations of the Anthropocene and other – cenes, as an epoch (for those working in the humanities) or as one of many geological events (Brannen 2019).

Climate change, by definition, includes measurements over at least thirty years and hence is long-term (IPCC 2013–2014; UNFCCC 1992). Similarly, in being parameterised by indicators such as global mean temperature (in the atmosphere and oceans) and concentration of specific gases in the atmosphere, it is by definition global-scale (IPCC 2013–2014; UNFCCC 1992). Downscaling climate models and attributing local meteorological phenomena to climate change has advanced impressively (National Academies of Sciences, Engineering, and Medicine 2016), but still remains challenging, again keeping climate change less integrated into personal, smaller, and shorter-term scales. Vulnerability assists in bridging this gap across scales and across environmental phenomena.

Process-based

First, in relation to temporality, vulnerability can be defined as a multi-scalar and ongoing process. Vulnerability takes into consideration the historical and structural processes – including policy decisions, omissions, and wider political contexts – which can range from lack of adequate building codes in hazard-exposed areas to further de-regulation or monoculture of already fragile economies that lead to further reproduction of socio-economic inequalities. These approaches formed the basis for empirical vulnerability analyses when the field was in its infancy. For example, authors such as Copans (1975) and Glantz (1976) analysed the 1968–1974 drought and famine in the Sahel by considering political processes including colonialism, neo-colonialism, sedenterisation, and marginalisation which meant that peoples who could previously cope with drought (e.g., Fleurett 1986), and thus were not really vulnerable to drought, were then made vulnerable to it. Irrespective of human-caused environmental changes, irrespective of the Anthropocene, the local and national social and political changes had created vulnerability to a changing environment. Similarly, Lewis (1984) selects the island of Antigua to conceptualise a historical and processual approach to explaining the environmental hazards which affect the location and hence how vulnerability to these hazards ought to be addressed.

While tying together physical and political (including economic) aspects, as well as their translation in and into lived experience, vulnerability as an ongoing and multi-scalar process directly challenges the simplifying and simplistic score-based vulnerability indexes (e.g., Cutter et al. 2000; EVI 2004). These indices position vulnerability as a score sum, sometimes weighted, of a range of indicators for selected jurisdictions – with indicators potentially being anything from earthquakes to population growth to tourism arrivals to violent conflict. In effect, they fall short of accounting for the extreme internal diversification of vulnerability, even within the same community, as well as its relational and ongoing character. Regardless of their potential value of giving a general sense of the complexity and inequality of global inequalities, these indices cannot possibly account for the rapidity and scale of impact of political changes and hazards (e.g., the 2018–2019 petrol crisis in the Caribbean and Latin America or the 2017 hurricane season across the Caribbean) as well as their long-term aftermaths, experienced as the long-running vulnerable present. Vulnerability as a process (Carrigan 2015; Kelman et al. 2015; Lewis 1999, 2009; Mika 2019) recognises and includes the uneven but essential interconnectedness, of people, places, environment, infrastructure, and services. Being a process, it is continuous, rather than going through distinct stages, phases, or cycles, thereby matching with conceptualisations of the disaster process (Carrigan 2015) and the ongoing and usually unfinished nature of post-disaster recovery (Mika 2019).

Comprehensive critical engagement

An anchoring in vulnerability necessitates an in-depth engagement with questions of ethics (e.g., Glantz 2003), responsibility (McEntire 2005), and the concomitant 'capacity or inclination to respond ethically to suffering at a distance'

(Butler 2012, 134). This is a necessary and urgently needed deepening of social and scientific considerations of vulnerability through including insights and long-standing discussions of vulnerability across history, the humanities, and other fields. Here, insights from critical analyses of 'bare life' (Agamben 1998), vulnerability and fear (Bauman 2006), or analytical distinctions between precarity and precariousness (Butler 2004, 2009), would greatly enrich ethical considerations of vulnerability, pushing these, in effect, beyond declarative commitments to 'cultural and local' differences in the language of policy documents and the ongoing 'neglect of integrating socio-cultural characteristics into DRR strategies' (Baytiyeh 2017, 63; Krüger et al. 2015).

The core concern at the heart of these respective analytical frames, if variously phrased is: what makes lives liveable? This vulnerability-orientated question drives beyond typical attempts to bypass history, previous understandings, and deeper frameworks. One example is climate change in particular (e.g., IPCC 2013–2014) forcing paradigms including 'resilience' and 'security' despite the heavy critiques of them. The critics of these approaches, such as Alexander (2013) for resilience and Chmutina et al. (2018) for security, base their analyses on foundational material which was available and known before the paradigms became prominent, yet the proponents of resilience, human security, and others do not factor in (or attempt to refute) the earlier literature. Conversely, the vulnerability process approach continues its critical engagement with past and ongoing work (e.g., Bankoff 2004) while also pushing forward theory and understanding (e.g., Lewis 2019) to continually seek improvement.

Furthermore, in no way does this expansive vulnerability-based framework bypass the necessity of undoing the coloniality of knowledge formations, building, instead on the understanding of the critiques of vulnerability as a Western discourse (Bankoff 2001; Bankoff et al. 2004) In fact, sustainable development including disaster risk reduction and climate change adaptation suffer similarly and legitimately, especially when these phrases, their definitions, and their international policy frameworks are limited to a top-down Western perspective, disguised as an 'international' approach. The challenge is to draw from that critique in order to capitalise on what the (renewed) term of 'vulnerability' could do while accepting and then overcoming its evident limitations.

Climate change adaptation integrated into disaster risk reduction

Revisiting and re-centring on original conceptualisations of vulnerability (e.g., Lewis 1979; O'Keefe et al. 1976) has two-fold, scholarly and political implications. First, it further contributes to meaningful integration of multiple scholarly perspectives, disciplines, approaches, and methodologies, namely by including and working with different methods rather than using merely the same, e.g., quantitative, methods in varying fields. Second, in extending the scope of analysis and action, an anchoring in vulnerability, rather than in exclusively climate change, the Anthropocene, or variations, joins both long- and short-term policy planning and practices while also maintaining a joint historical and future-orientated perspective that cannot be reduced to stages or cycles of progress or indicators.

As one example, climate change adaptation by definition (IPCC 2013–2014; UNFCCC 1992) deals with climate change only, which is defined as covering at least thirty years, hence its effects can be viewed only over the long-term, even for immediate and short-term actions which are part of it. Meanwhile, disaster risk reduction, by older and newer definitions (UNISDR 2009, 2017), embraces any activity relevant to its goal, irrespective of the time scale of the activity and its effects. Thus, disaster risk reduction covers the timescales of climate change adaptation and more. Similarly, climate change adaptation, by definition, addresses hazards and opportunities which might arise from the changing climate while disaster risk reduction covers all hazards and opportunities, whether from the atmosphere, hydrosphere/cryosphere, lithosphere, biosphere, or outer space. Again, the definitions mean that disaster risk reduction covers all of climate change adaptation and more.

Yet disaster risk reduction including climate change adaptation cannot cover all risks which people encounter. Many day-to-day risks can be ascribed to specific phenomena, such as road safety, tobacco, alcohol, harassment, and non-communicable diseases including obesity, while others are more about principles, ways, and possibilities of living, such as justice, equity, oppression, marginalisation, and discrimination. Disaster risk reduction including climate change adaptation typically recognises the principles and ways of living as potential underlying vulnerabilities but does not always incorporate the day-to-day risks. These everyday risks, occurring much more frequently than less common phenomena such as volcanic eruptions and ocean acidification, tend to dominate people's behaviour and concerns. A framework covering all these topics – that is, the gamut of society-environment interactions – is often articulated as development, sustainability, or sustainable development, leading to, for instance, the Sustainable Development Goals (UNGA 2015). A vulnerability perspective permits disaster risk reduction including climate change to sit within sustainable development, linking the wide variety of phenomena, scales, and issues which are of concern (Kelman 2017).

Vulnerability emphasises the need to work around simpler terms (which does not mean simplistic). This could mean dropping 'climate change adaptation' and 'disaster risk reduction' to use, instead, terms which are more readily understood across languages, cultures, and contexts. Some examples, which would not work in all specific situations and which have their own limitations, could be 'preventing disasters,' 'adjusting to hazards,' 'dealing with the environment,' or the most powerful 'tackling vulnerability.'

Limitations

Yet vulnerability does not and cannot cover all aspects of the difficulties which humanity has created for ourselves and the planet, whether or not this point refers to the Anthropocene or other '-cenes'/scenes. The word and concept of 'vulnerability' do not translate into many other languages or cultures, leading to similar limitations raised for other phrases in the previous section. Linguistically and conceptually, concerns are raised about mentioning merely 'vulnerability' without explaining what is vulnerable and to what it is vulnerable. Does 'vulnerability' have meaning without clarifying 'vulnerability to'?

In answering such questions, a balance is needed between individual and collective responsibilities. Focusing on only influences beyond an individual, or primarily systemic factors, might appear to absolve individuals of the responsibility to deal with their own vulnerabilities. In no way does this deny the experiences of the majority of individuals in being caught in traps of inequity, injustice, oppression, and marginalisation, leaving them with few opportunities to do better. Nor should people's own capabilities, knowledge, interests, and duties be downplayed, because almost everyone has something to contribute to vulnerability reduction and they should never hesitate in making this contribution. Without proper explanation or interpretation, an integrated, open-ended, encompassing vulnerability process might be seen as ignoring what individuals can and should contribute.

This can also be the case when using 'vulnerability' as a label from external perspectives. Describing individuals, groups, or locations as vulnerable due to long-term, long-standing processes beyond their control has the potential for sapping their will and agency to act or to change their own situation, immobilising the given community in a disempowering portrayal. Again, no claim is made that people can always enact the change they seek. But nor are most people or groups helpless victims, waiting for something to be done to them. Ultimately, balance is needed between being vulnerable and being made to be vulnerable – between being forced into situations and being able to control one's situation – with these balances manifesting in different ways for different contexts.

Vulnerable Anthropocenes

In sum, focusing the debates of climate change, anthropogenic environmental impacts, and related topics on vulnerability as a process provides, enhances, and supports:

1. Integration, by covering all topics, space scales, and time scales, while linking them.
2. Open-endedness, by seeking and expecting critical engagement and not bounding discussions by time or space scales.
3. Temporality, by being historical, present-focused, and future-orientated at all time and space scales, so neglecting neither short-term nor long-term aspects and including but going beyond prediction, projection, policy, and project/programme management.

This chapter, by interweaving perspectives from climate change policy, disaster studies and postcolonial and decolonial approaches in the humanities, has thus argued how vulnerability can offer a deep, broad, and processual analytical take bringing together near-comprehensive approaches to environmental and human changes, and their interactions, creating a space for conceptual and policy collaborations. Vulnerability, then, also encourages a more political and hence realistic entry point into and driver for debates on climate change, wider social and environmental concerns, social-environmental links, and how to act on the concerns raised. These entry points and actions are, in ways, more rigorous than the 'epoch debates', especially by acknowledging limitations (e.g., cultural and linguistic translation and interpretation) while not being confined through artificial, definition-related, threshold-based boundaries.

In fact, thresholds and boundaries of parameters said to delineate acceptable regimes from unacceptable ones (e.g., planetary boundaries) are explicitly avoided as being too mechanistic and too definitive. From climate change, many discussions have pointed out how specific levels of parts per million in the atmosphere (namely of CO_2-equivalent) or specific temperature rises (namely of degrees Celsius above the pre-industrial global mean temperature) are political rather than scientific selections. There is no question about rejecting the idea that 1.99°C means a wonderful world whilst 2.01°C means catastrophe. Nor does this statement mean that 2°C (or 2.00°C) is a pointless political target. To a large extent, the difficulty is in quantification itself alongside the desire and expectation to quantify, both of which lend undeserved credibility to numbers. This is the case even when the numbers help to bring people on board through definite statements such as '1.5°C to stay alive' (Deitelhoff and Wallbott 2012) and 350.org (McKibben 2013), because it leaves such initiatives wide open to legitimate scientific critique and hence political disparaging.

Vulnerability – while not being immune from ideological, political, and scientific challenges, and while not covering every aspect perfectly – provides a much wider and deeper methodological and analytical approach, generally steeped in a combination of knowledge forms and histories, disciplines, and professions, allowing still to foreground the very questions of knowledge, power and expertise formation, inherent to all of them. Vulnerability does not deny the urgency and complexity of the current challenges to unjust systems that render lives less liveable, instead highlighting these characteristics within the long-standing contexts which bring climate change forth as one human-caused social and environmental challenge among many. It also provides an anchor which brings this 'many' together in order to formulate actions and solutions which tackle the fundaments of the causes and might not create further problems. Consequently, if embraced appropriately, vulnerability can circumvent the need for quantification and the inherent bias therein (Martin 1979). The integration, open-endedness, and temporality lend themselves to multi-scalar and processual approaches helping to address uneven experiences of environmental pasts, presents, and futures. It thereby embraces but moves beyond various conceptualisations of the Anthropocenes, the vulnerabilities expressed by them, and the vulnerabilities of the concept itself.

References

Agamben, Giorgio. 1998. *Homo Sacer: Sovereign Power and Bare Life*. Stanford: Stanford University Press.

Alexander, David E. 2013. "Resilience and Disaster Risk Reduction: An Etymological Journey." *Natural Hazards and Earth Systems Sciences* 13, no. 11: 2707–2716.

Bankoff, Greg. 2001. "Rendering the World Unsafe: 'Vulnerability' as Western Discourse." *Disasters* 25, no: 1: 19–35.

Bankoff, Greg. 2004. "Time is of the Essence: Disasters, Vulnerability and History." *International Journal of Mass Emergencies and Disasters* 22, no. 3: 23–42.

Bankoff, Greg, Georg Frerks, and Dorothea Hilhorst, eds. 2004. *Mapping Vulnerability: Disasters, Development and People*. London: Earthscan.

Bauman, Zygmunt. 2006. *Liquid Fear*. Cambridge: Polity Press.

Baytiyeh, Hoda. 2017. "Socio-cultural Characteristics: The Missing Factor in Disaster Risk Reduction Strategy in Sectarian Divided Societies." *International Journal of Disaster Risk Reduction* 21: 63–69.

Brannen, Peter. 2019. "The Anthropocene Is a Joke: On Geological Timescales, Human Civilization Is an Event, Not an Epoch." *The Atlantic*, August 13, 2019. https://www.theatlantic.com/science/archive/2019/08/arrogance-anthropocene/595795.

Butler, Judith. 2004. *Precarious Life: The Powers of Mourning and Violence*. London: Verso.

Butler, Judith. 2009. *Frames of War: When Is Life Grievable?* London: Verso.

Butler, Judith. 2012. "Precarious Life, Vulnerability, and the Ethics of Cohabitation." *The Journal of Speculative Philosophy* 26, no. 2: 134–151.

Cannon, Terry. 1994. "Vulnerability Analysis and The Explanation of 'Natural Disasters'." In *Disasters, Development and Environment*, edited by Ann Varley, 13–30. Chichester: John Wiley & Sons.

Carrigan, Anthony. 2015. "Towards a Post-Colonial Disaster Studies." In *Global Ecologies and the Environmental Humanities: Postcolonial Approaches*, edited by Elizabeth Deloughrey, Jill Didur, and Anthony Carrigan, 117–138, Abingdon: Routledge.

Chakrabarty, Dipesh. 2000. *Provincializing Europe: Postcolonial Thought and Historical Difference* (New ed). Princeton: Princeton University Press.

Chakrabarty, Dipesh 2009. "The Climate of History: Four Theses." *Critical Inquiry* 35, no. 2: 197–222.

Chester, David K. 1993. *Volcanoes and Society*. London: Edward Arnold.

Chmutina, Ksenia, Pete Fussey, Andrew Dainty, and Lee Bosher. 2018. "Implications of Transforming Climate Change Risks Into Security Risks." *Disaster Prevention and Management* 27, no. 5: 460–477.

Clark, Peter U., Jeremy D. Shakun, Shaun A. Marcott, Alan C. Mix, Michael Eby, Scott Kulp, Anders Levermann, Glenn A. Milne, Patrik L. Pfister, Benjamin D. Santer, Daniel P. Schrag, Susan Solomon, Thomas F. Stocker, Benjamin H. Strauss, Andrew J. Weaver, Ricarda Winkelmann, David Archer, Edouard Bard, Aaron Goldner, Kurt Lambeck, Raymond T. Pierrehumbert, and Gian-Kasper Plattner. 2016. "Consequences of twenty-first-century policy for multi-millennial climate and sea-level change." *Nature Climate Change* 6: 360–369.

Copans, Jean, ed. 1975. *Sécheresses et famines du Sahel*. Paris: F. Maspero.

Cutter, Susan L., Jerry T. Mitchell, and Michael S. Scott. 2000. "Revealing the Vulnerability of People and Places: A Case Study of Georgetown County, South Carolina." *Annals of the Association of American Geographers* 90, no. 4: 713–737.

Crutzen, Paul J. and Eugene F. Stoermer. 2000. "The Anthropocene." *Global Change Newsletter* 41: 17–18.

Crutzen, Paul J. 2002. "Geology of Mankind." *Nature* 415: 23.

Davies, Jeremy. 2016. *The Birth of the Anthropocene.* Oakland: University of California Press.

Deitelhoff, Nicole, and Linda Wallbott. 2012. "Small States in International Negotiations Beyond Soft Balancing: Small States and Coalition-Building in the ICC and Climate Negotiations." *Cambridge Review of International Affairs* 25, no. 3: 345–366.

DeLoughrey, Elizabeth. 2019. *Allegories of the Anthropocene.* Durham: Duke University Press.

EVI. 2004. "The Environmental Vulnerability Index (EVI) 2004." *SOPAC Technical Report 384.* Suva: SOPAC (South Pacific Applied Geoscience Commission).

Fleurett, Anne. 1986. "Indigenous Responses to Drought in Sub-Saharan Africa." *Disasters* 10, no. 3: 224–229.

Gaillard, JC. 2019. "Disaster Studies Inside Out." *Disasters* 4, no. S1: S7–S17.

Ginn, Franklin, Michelle Bastian, David Farrier, and Jeremy Kidwell. 2018. "Introduction: Unexpected Encounters with Deep Time." *Environmental Humanities* 10, no. 1: 213–225.

Glantz, Michael. 1976. *The Politics of Natural Disaster: The Case of the Sahel Drought.* New York: Praeger.

Glantz, Michael. 2003. *Climate Affairs: A Primer.* Covelo: Island Press.

Haraway, Donna. 2015. "Anthropocene, Capitalocene, Plantationocene, Chthulucene: Making Kin." *Environmental Humanities* 6, no. 1: 159–165.

Haraway, Donna. 2016. "Staying with the Trouble." In: *Anthropocene or Capitalocene?*, edited by Jason W. Moore, 34–76, Oakland: PM Press.

Haraway, Donna, Noboru Ishikawa, Scott F. Gilbert, Kenneth Olwig, Anna L. Tsing, and Nils Bubandt. 2015. "Anthropologists Are Talking – About the Anthropocene." *Ethnos* 81, no. 3: 535–564.

Hewitt, Kenneth, ed. 1983. *Interpretations of Calamity.* London: Allen & Unwin.

Hewitt, Kenneth. 1997. *Regions of Risk: A Geographical Introduction to Disasters.* London: Routledge.

IPCC. 2013–2014. *IPCC Fifth Assessment Report.* Geneva: IPCC (Intergovernmental Panel on Climate Change).

Kelman, Ilan. 2017. "Linking Disaster Risk Reduction, Climate Change, and the Sustainable Development Goals." *Disaster Prevention and Management* 26, no. 3: 254–258.

Kelman, Ilan, J. C. Gaillard, and Jessica Mercer. 2015. "Climate Change's Role in Disaster Risk Reduction's Future: Beyond Vulnerability and Resilience." *International Journal of Disaster Risk Science* 6, no. 1: 21–27.

Knutson, Thomas, Suzana J. Camargo, Johnny C. L. Chan, Kerry Emanuel, Chang-Hoi Ho, James Kossin, Mrutyunjay Mohapatra, Masaki Satoh, Masato Sugi, Kevin Walsh, and Liguang Wu. 2020. "Tropical Cyclones and Climate Change Assessment: Part II: Projected Response to Anthropogenic Warming." *Bulletin of the American Meteorological Society* 101, no. 3: E303–E322.

Krüger, Fred, Greg Bankoff, Terry Cannon, Benedikt Orlowski, and E. Lisa F. Schipper, eds. 2015. *Cultures and Disasters: Understanding Cultural Framings in Disaster Risk Reduction.* Abingdon: Routledge.

Lewis, James. 1979. "The Vulnerable State: An Alternative View." In *Disaster Assistance: Appraisal, Reform and New Approaches*, edited by Lynn Stephens and Stephen J. Green, 104–129, New York: New York University Press.

Lewis, James. 1984. "A multi-Hazard History of Antigua." *Disasters* 8, no. 3, 190–197.

Lewis, James. 1999. *Development in Disaster-Prone Places: Studies of Vulnerability*. London: Intermediate Technology Publications.

Lewis, James. 2009. "An Island Characteristic: Derivative Vulnerabilities to Indigenous and Exogenous Hazards." *Shima* 3 no. 1: 3–15.

Lewis, James. 2019. "The Fluidity of Risk: Variable Vulnerabilities and Uncertainties of Behavioural Response to Natural and Technological Hazards." *Disaster Prevention and Management* 28, no. 5: 636–648.

Malm, Andreas and Alf Hornborg. 2014. "The Geology of Mankind? A Critique of the Anthropocene Narrative." *The Anthropocene Review* 1, no. 1: 62–69.

Malm, Andreas. 2016. *Fossil Capital*. London: Verso.

Martin, Brian. 1979. *The Bias of Science*. Canberra: Society for Social Responsibility in Science.

McEntire, David A. 2005. "Why Vulnerability Matters." *Disaster Prevention and Management* 14, no. 2: 206–222.

McKibben, Bill. 2013. "Don't Imagine the Future – It's Already Here." *Organization* 20, no. 5: 745–747.

Mignolo, Walter D. 2008. "Coloniality: The Darker Side of Modernity." In *Coloniality at Large: Latin America and the Postcolonial Debate* edited by Mabel Moraña, Enrique D. Dussel, and Carlos A. Jáuregui, 39–49, Durham NC: Duke University Press.

Mignolo, Walter D. 2011. *The Darker Side of Western Modernity: Global Futures, Decolonial Options*. Durham NC: Duke University Press.

Mignolo, Walter D. 2013. "Geopolitics of Sensing and Knowing: On (de)coloniality, Border Thinking, and Epistemic Disobedience." *Confero: Essays on Education, Philosophy and Politics* 1, no. 1: 129–150.

Mika, Kasia. 2019. *Disasters, Vulnerability, and Narratives: Writing Haiti's Futures*. Abingdon: Routledge.

Moore, Jason W. 2017. "The Capitalocene, Part I: On the Nature and Origins of Our Ecological Crisis." *The Journal of Peasant Studies* 44, no. 3: 594–630.

Moore, Jason W. 2018. "The Capitalocene Part II: Accumulation by Appropriation and the Centrality of Unpaid Work/energy." *The Journal of Peasant Studies* 45, no. 2: 237–279.

Monsaingeon, Baptiste. 2017. *Homo Detritus: Critique de la société du déchet*. Paris: Seuil.

National Academies of Sciences, Engineering, and Medicine. 2016. *Attribution of Extreme Weather Events in the Context of Climate Change*. Washington, DC: National Academies Press.

O'Keefe, Phil, Ken Westgate, and Ben Wisner. 1976. "Taking the Naturalness Out of Natural Disasters." *Nature* 260: 566–567.

Reed, Christina. 2015. "Plastic Age: How it's Reshaping Rocks, Oceans and Life." *New Scientist*, January 29, 2015. https://www.newscientist.com/article/mg22530060-200-plastic-age-how-its-reshaping-rocks-oceans-and-life

Romero, Romualdo, and Kerry Emanuel. 2017. "Climate Change and Hurricane-Like Extratropical Cyclones: Projections for North Atlantic Polar Lows and Medicanes Based on CMIP5 Models." *Journal of Climate* 30, no. 1, 279–299.

Shepherd, Nick. 2018. "Decolonial Thinking & Practice." *ECHOES: European Colonial Heritage Modalities in Entangled Cities*. https://keywordsechoes.com/decolonial-thinking-and-practice.

Torry, William I. 1979. "Hazards, Hazes And Holes: A Critique of The Environment as Hazard and General Reflections On Disaster Research." *Canadian Geographer* 23, no. 4: 368–383.

UN. 2000. *Millennium Development Goals*. New York: UN (United Nations).

UNFCCC. 1992. *United Nations Framework Convention on Climate Change*. Bonn: UNFCCC (United Nations Framework Convention on Climate Change).

UNFCCC. 1997. *Kyoto Protocol to the United Nations Framework Convention on Climate Change*. Bonn: UNFCCC (United Nations Framework Convention on Climate Change).

UNFCCC. 2015. *The Paris Agreement*. Bonn: UNFCCC (United Nations Framework Convention on Climate Change).

UNGA. 2015. *Resolution Adopted by the General Assembly on 25 September 2015, A/RES/70/1*. New York: UNGA (United Nations General Assembly).

UNISDR. 2005. *Hyogo Framework for Action 2005–2015: Building the Resilience of Nations and Communities to Disasters*. Geneva: UNISDR (United Nations International Strategy for Disaster Reduction).

UNISDR (United Nations International Strategy for Disaster Reduction). 2015. *Sendai Framework for Disaster Risk Reduction 2015–2030*. Geneva: UNISDR (United Nations International Strategy for Disaster Reduction).

UNISDR. 2009. *Terminology*. Geneva: UNISDR (United Nations International Strategy for Disaster Reduction).

UNISDR. 2017. *Terminology on Disaster Risk Reduction*. Geneva: UNISDR (United Nations International Strategy for Disaster Reduction).

Watts, Nick, Markus Amann, Nigel Arnell, Sonja Ayeb-Karlsson, Jessica Beagley, Kristine Belesova, Maxwell Boykoff, Peter Byass, Wenjia Cai, Diarmid Campbell-Lendrum, Stuart Capstick, Jonathan Chambers, Samantha Coleman, Carole Dalin, Meaghan Daly, Niheer Dasandi, Shouro Dasgupta, Michael Davies, Claudia Di Napoli, Paula Dominguez-Salas, Paul Drummond, Robert Dubrow, Kristie L Ebi, Matthew Eckelman, Paul Ekins, Luis E Escobar, Lucien Georgeson, Su Golder, Delia Grace, Hilary Graham, Paul Haggar, Ian Hamilton, Stella Hartinger, Jeremy Hess, Shih-Che Hsu, Nick Hughes, Slava Jankin Mikhaylov, Marcia P Jimenez, Ilan Kelman, Harry Kennard, Gregor Kiesewetter, Patrick L Kinney, Tord Kjellstrom, Dominic Kniveton, Pete Lampard, Bruno Lemke, Yang Liu, Zhao Liu, Melissa Lott, Rachel Lowe, Jaime Martinez-Urtaza, Mark Maslin, Lucy McAllister, Alice McGushin, Celia McMichael, James Milner, Maziar Moradi-Lakeh, Karyn Morrissey, Simon Munzert, Kris A Murray, Tara Neville, Maria Nilsson, Maquins Odhiambo Sewe, Tadj Oreszczyn, Matthias Otto, Fereidoon Owfi, Olivia Pearman, David Pencheon, Ruth Quinn, Mahnaz Rabbaniha, Elizabeth Robinson, Joacim Rocklöv, Marina Romanello, Jan C Semenza, Jodi Sherman, Liuhua Shi, Marco Springmann, Meisam Tabatabaei, Jonathon Taylor, Joaquin Triñanes, Joy Shumake-Guillemot, Bryan Vu, Paul Wilkinson, Matthew Winning, Peng Gong, Hugh Montgomery, and Anthony Costello. 2020. "The 2020 Report of The *Lancet* Countdown on Health and Climate Change: Responding to Converging Crises." *The Lancet* 397, no. 10269: 129–170.

Usher, Phillip John. 2016. "Untranslating the Anthropocene." *Diacritics* 44, no. 3: 56–77.

Wisner, Ben. 1993. "Disaster Vulnerability: Scale, Power and Daily Life." *GeoJournal* 30, 127–140.

Wisner, Ben, Piers Blaikie, Terry Cannon, and Ian Davis. 2004. *At Risk: Natural Hazards, People's Vulnerability and Disasters*, 2nd ed. London: Routledge.

13

'THE HOTTEST SUMMER EVER!'

Exploring vulnerability to climate change among grain producers in Eastern Norway

Bjørnar Sæther and Karen O'Brien

Introduction: the changing landscape of vulnerability

'What makes people vulnerable?' This question, posed by Hilhorst and Bankoff (2004, 1) in the introduction to *Mapping Vulnerability: Disasters, Development and People*, is a critical one to consider if we are interested in a sustainable and resilient future. Although there is a logic and appeal to the current focus on adaptation and resilience, without an understanding of the root causes and drivers of vulnerability it becomes easy to promote resilience and adaptation within the very systems that are contributing to vulnerability in the first place. Understanding the multiple dimensions and dynamics of vulnerability is a prerequisite for transformative responses to risk reduction.

Hilhorst and Bankoff (2004) offered a nuanced answer to the question of what makes people vulnerable, pointing out that at one level it is about poverty, resource depletion, and marginalisation, and at another level about the diversity of risks generated by both local and global processes. Alluding to the changing nature of disasters, they called for attention to the dynamic interactions and linkages that generate destructive forces. They also reminded us that vulnerability is about people and their ideas, perceptions, and practices in relation to risk and disaster (Hilhorst and Bankoff 2004). To reduce vulnerability in practice, issues of empowerment, capacity, local participation, and organisational strengthening were recognised as vital (Frerks and Bender 2004).

More than 15 years after the publication of *Mapping Vulnerability*, the question of what makes people vulnerable must be examined within the context of accelerating global change. Climate change is shattering temperature records, disrupting rainfall patterns, and normalising extreme weather events, and there are increasing concerns about feedbacks in the climate system that could trigger tipping points leading to a 'Hothouse Earth' scenario (Steffen et al. 2018).

DOI: 10.4324/9781003219453-16

Environmental changes have been transforming vulnerability landscapes for many households, communities, sectors, and social groups. Despite significant attention to strategies and actions to promote adaptation and resilience, millions of people experience vulnerabilities linked to multiple and interacting stressors, including those related to extreme weather events and climate-related hazards (Leichenko and O´Brien 2019).

The vulnerability landscape is becoming deeper and more extensive than many experts anticipated two or three decades ago, and we are moving into unknown territory. For example, an increase in bushfires in Australia has revealed that there is little knowledge on which subgroups of the population are most vulnerable to the long-term effects of smoke (Yu et al. 2020). Increasing exposure to heatwaves in European capitals may impose greater risks for urban residents who are elderly, young, isolated, or with pre-existing chronic conditions or mental health challenges, as well as communities with weak socioeconomic status (Smid et al. 2019). Case after case reveals that the dynamics of vulnerability are closely tied to economic, political, and social challenges, including a growing concentration of wealth, dissatisfaction with and a backlash to current forms of democracy, economic uncertainty, and the risks associated with global pandemics (Piketty 2020; Ribot 2014).

Many groups and communities that have felt complacent about their resilience and adaptive capacity are increasingly confronted with climate change impacts that are intertwined with other social, economic, and ecological dynamics. There is a growing awareness that actions must be taken to reduce risk to multiple, interacting stressors. Yet is there any evidence that approaches to vulnerability reduction are transforming as quickly as the vulnerability landscape? To engage with this inquiry, we explore what vulnerability looks like and how it is perceived by those who generally consider themselves to be resilient to variations and extremes in weather and climate, namely farmers in Eastern Norway.

Norway is a country that is widely considered to be resilient to climate change based on its wealth, education, infrastructure, access to resources, and management capabilities, i.e., the factors contributing to high adaptive capacity at the national level (O'Brien et al. 2004; Sarkodie and Strezov 2019). However, such conceptualisations of resilience are dependent on perceptions of risk, vulnerability, and adaptive capacity, which are often challenged when vulnerability is seen through an integrated, multi-scale lens (O'Brien et al. 2006; Slovic 2000). The extremely hot summer of 2018 in Norway provides an opportunity to study the current discourse on vulnerability among farmers in Eastern Norway, a region that accounts for 80 per cent of the country's grain production. The 'hottest summer ever' was experienced in a region that has seen average temperature increases of about 1°C over the past 30 years. In 2018, the drop in production of grain and grass by about 50 per cent led to a dramatic reduction in farm-level income, and a wide range of consequences and responses. In this chapter, we investigate the following:

- What were the consequences of the drought at the farm level?
- What responses can be identified at the farm level?
- How did the institutions in agriculture respond to the drought?

The implications of different responses will be discussed in light of the transformative changes needed to reduce risk and vulnerability across local, national, and global scales.

We begin by describing the 'hottest summer ever' and its impacts on Norwegian agriculture. We then draw on interviews with farmers about their ideas, perceptions, and practices related to vulnerability in the aftermath of the summer of 2018, with attention to the uneven impacts and differential responses among grain farmers in Eastern Norway. We relate these responses to the practical, political, and personal spheres of transformations (O'Brien 2018). Our results show that most of the actions taken by farmers to reduce vulnerability involved technical and behavioural responses that fall within the practical sphere. Yet in many cases, these responses were facilitated by structures and systems related to the political sphere of transformation, including cultural norms and formal and informal institutions. However, one important driver of transformative change for reducing long-term global vulnerability was notably absent: some farmers do not believe that climate change is a risk, and few linked vulnerability of agriculture to the perpetuation of an oil-based, consumption-oriented economy that is contributing to increased risk and vulnerability at a global scale. A sense of complacency supports technical responses over deeper and more extensive transformations that could address the long-term vulnerability of farmers in Eastern Norway and elsewhere in the world.

We draw on an extreme case in a wealthy country that is considered 'resilient' to climate change because it provides an important perspective on the evolving vulnerability landscape. Moreover, it can also offer important insights on the types of transformations that may be needed to reduce vulnerability. Our study makes use of data on agricultural production and applications for public drought relief available from Statistics Norway, together with weather and climate data from the Norwegian Meteorological Institute. However, there are no statistics available on how farmers coped with the drought and possible strategies to prepare for future climate extremes. To understand how vulnerability to climate change was experienced by farmers, we interviewed a limited number of farmers and read about 100 newspaper articles on the drought. In addition, one of the authors operates a small grain farm in Eastern Norway and directly experienced the drought and its consequences. He took part in formal and informal conversations about the drought with neighbours and in the local farmers association. By combining insights based on facts on the ground in Eastern Norway with discussions of vulnerability to climate change and extreme weather events, this theoretically-informed case study seeks to contribute to theory development (George and Bennett 2005; Ragin 1994). In studies of single cases many instances can be investigated, and the interplay between evidence-based images and theoretical ideas can lead to progressive refinement of both through a process of retroduction (Ragin 1994). Theory development and refinement of key concepts seems to be increasingly relevant in a world likely to experience even more severe climate change in the decades ahead (Field et al. 2012).

'The hottest summer ever' in Norway

Climate change will have significant consequences for Norwegian agriculture. According to some climate models, the growing season in Norway could increase by one month by 2050, and by two months towards the end of this century (Hanssen-Bauer et al. 2017). Earlier research suggested that climate change could be beneficial for Norwegian agriculture, with grain production expected to increase by more than 50 per cent compared to 1992–1993, or more if multi-cropping and irrigation are included (Fischer et al. 2001). However, this positive prognosis does not take into consideration how the quantity and quality of yields might be influenced by increased incidents of pests and diseases, by soil erosion and nutrient deficiencies resulting from climate change, and by extreme weather events (Bechmann and Deelstra 2013; O'Brien et al. 2004).

The growing season has already become longer in Norway. However, rainfall has also increased by 20 per cent at the national level. This has led to an increase in the number of incidents of heavy rainfall. Norway has experienced several summers and autumns with very wet conditions, leading to crop losses and increased costs and reduced income for farmers. There is growing concern over the long-term effects of wetter conditions, as this contributes to increased runoff, nutrient loss, and soil erosion, as well as soil compaction by heavy machinery. This has already resulted in reduced yield per hectare for many farmers (Bardalen et al. 2018).

Adapting to warmer and wetter conditions is a challenge for Norwegian farmers. There are some signs that farmers have already started to adapt to wetter conditions, for example by increasing investments in constructing ditches to remove excess water. Such measures will improve plant growth and reduce emissions of CO_2 and nitrous gases (Norwegian Environment Agency 2020). This work is, however, in an early phase and knowledge about the links between agricultural practices and emissions of CO_2, methane and other gases is still limited (Bardalen et al. 2018).

However, what happens when the problem is not too much rainfall, but too little? How does this affect the vulnerability of farmers, and of the agricultural sector more generally? A new landscape of vulnerability became evident during the summer of 2018, when temperatures records were broken in many locations in Europe north of the Alps, contributing to extreme impacts such as wildfires, drought, and heatwaves (Buraas et al. 2020). The hot and dry conditions seen in Europe north of the Alps in 2018 were experienced as record-breaking temperatures in Norway.

According to the Norwegian Meteorological Institute, the very warm May-July period in Norway in 2018 was remarkable, with temperatures on average 1.2°C warmer than the previous record in 2002 (Gangstø Skarland et al. 2019). The period from May to July had an average temperature 3.1°C above normal and 74 per cent of normal precipitation. Nationally, it was the fourth driest May to July period since data were first recorded, starting in the year 1900 (Gangstø Skaland et al. 2019). The combination of extremely high temperatures and well below average rainfall had considerable consequences for agriculture, and little drought adaptation of ecosystems in Norway has contributed to this (Buraas et al. 2020).

According to Statistics Norway, the decline in agricultural output in 2018 lowered mainland economic growth by 0.2 per cent (Bougroug and Ånestad 2019). Abnormally warm temperatures throughout the summer resulted in poor plant growth and forced maturation. The fact that the drought started in May contributed to this (Buraas et al. 2020). However, these physical impacts were experienced within a changing socio-economic environment. Agriculture in Norway has been undergoing structural change, and currently consists of 39,000 farmers operating an average farm size of 25 hectares. Since 1990 the number of farmers has been reduced by 60 per cent and the average farm size has increased by 150 per cent, with more farmland concentrated in the hands of fewer farmers. Only three per cent of Norway's land is suited to agricultural production, and about one per cent is suitable for growing grain for human consumption. The agricultural sector in Norway is small compared to neighbouring countries Sweden and Denmark, and grain production is limited. As an example, Danish grain production is 6–8 times larger than in Norway (Statistics Denmark 2019).

Vulnerability in Eastern Norway

What was remarkable about the summer of 2018 in Eastern Norway, according to the Norwegian Meteorological Institute, was a mean temperature from May to July that was 4.3°C above the 1961–1990 average, combined with a 40 per cent reduction in rainfall (Gangstø Skaland et al. 2019, 17). The combination of very warm and dry weather contributed to very high evapotranspiration from the fields and the plants, which resulted in a particularly severe situation of agricultural drought. To illustrate the magnitude of the temperature anomalies, Table 13.1 presents temperature and precipitation data for the 2018 growing season (May–September) in the town of Årnes, located in a major grain production area in Eastern Norway. The maximum temperature reached 34.1°C in July, and there were seven days with maximum temperatures above 30 degrees. The low precipitation at Årnes, well below the average of Eastern Norway, illustrates local variation.

By early June, it was already evident to farmers that their fields were affected by the drought. The hardest-hit farmers produced yields in the range of 20 per cent of normal. At farms with less clay and more organic material in the topsoil, germination was more successful and fields were green by early June. However, the continuation of dry and warm conditions throughout the summer resulted in yields well below average for farmers in Eastern Norway. In terms of grain and grass production, yields in 2018 were reduced by 50 per cent compared to average yields (Statistics Norway 2019). Table 13.1 presents production data for two major grains; wheat and oats. The figures for wheat include both winter and spring varieties, the winter varieties were less hard hit than the spring varieties. Wheat production includes wheat for human consumption and fodder for animals (Table 13.2).

TABLE 13.1 Weather in the growing season at Årnes 2018

Month	*Temperature in °C*			*Precipitation in mm*		
	2018	*Average (1961–1990)*	*Difference*	*2018*	*Average (1961–1990)*	*Difference*
May	15.3	9.8	5.5	27.8	53	−25.2 (−47.5%)
June	16.3	14.2	2.1	36.8	60	−23.2 (−38.0%)
July	20.7	15.2	5.5	17.8	62	−44.2 (−71.3%)
August	14.8	14	0.8	43.2	86	−42.8 (−49.7%)
September	11.4	9.7	1.7	84.1	76	+8.1 (10.7%)

Source: Norwegian Meteorological Institute

TABLE 13.2 Production of wheat and oats in Eastern Norway in tons

Grain	*Average (2012–2016)*	*2018*	*2018 compared to 2012–2016 (%)*
Wheat	3,43,000	1,27,000	−63
Oats	2,56,000	1,44,000	−42

Source: Statistics Norway

While all farmers in Eastern Norway were affected by the drought, some farmers proved more vulnerable than others. Both soil conditions and farming practices influenced outcomes at the level of individual farms. For farms with topsoil with high clay content and little organic material, the problems started immediately after planting in early May, which is a critical time for seed germination. With temperatures as much as 6.5°C above normal, together with only half the normal amount of rainfall, many seeds were simply unable to sprout. This was particularly a problem in fields that had been plowed. When fields are plowed, soils dry up quickly in spring, which facilitates early planting. Under normal conditions, i.e., with normal temperatures and rainfall, this practice results in good germination in May and fields show the potential for a good harvest. As a consequence of the weather conditions in May and June, water content in the topsoil was too low (below six per cent) to allow for germination. This had dramatic consequences for crop yields.

A 50 per cent reduction in production resulted in a 50 per cent reduction in income from grain sales. The 'hottest summer ever' thus demonstrated that grain producers are economically vulnerable to extreme dry springs and summers. This vulnerability raises important questions concerning responses at both the level of individual farms and the societal level.

Responding to extremes: implications for the vulnerability landscape

Farmers and agricultural institutions in Norway responded to the warmer and drier summer of 2018 in a number of ways. Whether these responses address long-term risk and vulnerability, however, depends on whether 2018 is considered to be an extreme anomaly, or a harbinger of future conditions for farming in Norway. To structure our analysis of the range of responses discussed in newspapers and interviews, we consider how they correspond to three interacting spheres of transformation. This framework provides a heuristic for understanding relationships among the practical, political, and personal spheres of transformation that are involved in reducing vulnerability to climate extremes (O'Brien 2018). The practical sphere includes technical and behavioural responses that directly address the consequences of extreme weather. The political sphere represents changes to the systems and structures that facilitate or hinder actions in the practical sphere. The personal sphere includes the individual and shared beliefs, values, worldviews and paradigms that not only shape goals and outcomes but also influence social and cultural norms, how systems are organised, and how institutions relate to vulnerability and climate change risks.

Vulnerability reduction in the practical sphere

Vulnerability to drought in Eastern Norway can be linked to the characteristics of individual farmers, including their economic situation, amount of debt, and total household income that comes from farming activities. Income diversification was one factor that significantly reduced vulnerability to the 2018 drought. Multiple sources of income, including income from other activities than farming, reduced the economic consequences of the drought at the farm level. Only about 12 per cent of Norwegian farmers receive more than 90 per cent of their income from farming alone. Among farmers that produce only grain, no more than 1 per cent receive more than 90 per cent of income from farming.

Among the very few farmers who maintained normal yields were those who had irrigation equipment. Such equipment is generally considered too costly for grain production. Irrigation of large areas of farmland also requires stable access to large quantities of water. During the drought, many smaller rivers almost dried up, and only the largest rivers and lakes could support irrigation. The few farmers with irrigation equipment had to work around the clock to operate the machinery throughout the summer. This resulted in high costs related to both labour and energy. Nonetheless, they were able to harvest normal amounts of grain in August.

Due to the drought in 2018 the seed supplies in front of the 2019 season were critically low. In earlier years, wet summers have caused problems for the production of seeds, as heavy rains in August and September have a negative influence of maturing of grains and reduce the overall quality and the germination of the grain when it is planted the next spring. A series of wet seasons between 2015 and 2017 resulted in a large supply of seeds of rather low quality. In contrast, 'the hottest summer ever' of 2018 resulted in seeds of good quality, but in very limited supply. One direct consequence was limited availability of seeds for planting in the spring of 2019. The manager of one of the major suppliers of seeds warned farmers in a written statement in January 2019:

> The 2018 season resulted in seeds with good germination, but the harvest left us with limited supplies. Sometimes during the winter season we will be sold out of seeds grown in Norway. Due to our good relations with actors in neighboring countries we will be able to import seeds that to some degree have been tested in Norway. This import will help us cover most of the needs the coming spring.

Varieties of seeds used in major grain-growing countries like Germany and France are not suited for Norway, since the growing season in Central Europe is both longer and warmer. It is only Finland and mid-Sweden that have a growing season similar to Eastern Norway, thus the number of seed suppliers in this region is limited. Although major actors, including government institutions, are aware of the vulnerability of seed production in northern Europe, effective measures have yet to be taken to address this vulnerability. The poor harvest of 2018 accentuated questions concerning self-sufficiency due to limited supplies of domestic grain as a basic foodstuff and access to seeds for the next planting season. Access to seeds suitable for a changing climate in the north of Europe is a fundamental question for the future of farming.

Vulnerability reduction in the political sphere

The economic losses experienced in 2018 were distributed unequally among farmers. However, both cultural and institutional factors play a key role in reducing vulnerability. To understand the changing landscape of vulnerability in Norway, it is necessary to explore how the drought was handled in the political sphere, which relates to the cultural norms, institutions, regulations, and incentives that influence or impede practical responses. In the case of Eastern Norway, resources were mobilised across networks, regions, and institutions when the extent of the drought became clear in early July.

Cultural expressions of solidarity helped to reduce vulnerability to the drought. In a study of adaptation among farming communities in Norway, Eriksen and Selboe (2012) emphasise the importance of social relations and trust. In Norway, the attitude of a 'dugnad' or collective effort has a long tradition, especially during times of crisis. Indeed, mechanisms of trust and collective thinking were operating among farmers during 'the hottest summer ever.' Dairy farmers were hard hit by lack of fodder due to very small harvests of grass. Grain farmers were asked to collect straw from the harvest and offer it to dairy farmers as emergency fodder. As told by one informant:

> Contact was established between buyers and sellers of straw. (...) All the straw from the grain fields was collected and sold or given away to livestock farmers. It was an amazing 'dugnad' by the farmers and shows the solidarity in the sector. Farmers wanted to help each other.

The exchange of limited resources outside normal market mechanisms was organised by networks of farmers and local farmers associations. Volunteers set up a website to connect those with extra fodder with those who were in need. These efforts reduced vulnerability among dairy farmers.

The largest purchaser of grains and provider of seeds and other farm inputs in Norway is the cooperative 'Felleskjøpet,' owned by farmers. The cooperative mobilised its economic and organisational resources to help its owners through the drought. Due to the drought situation north of the Alps (Buraas et al. 2020), it was not possible to import fodder from nearby countries such as Denmark or Sweden. Canada offered to sell grass, but the offer was declined due to a high risk of importing unwanted species. Grass was instead bought from Iceland and distributed by Felleskjøpet. The cooperative also collaborated with the dairy farmers' cooperative, 'Tine,' to provide farmers with information about where and how to access fodder for animals. Norway's culture of cooperation, organised through formal and informal institutions, thus played a significant role in reducing vulnerability during 'the hottest summer ever.'

The institutional set-up of Norwegian agriculture also played a prominent role in supporting grain farmers in times of crisis. Most farmers are members of the Farmers Association and there are local branches in every municipality. During the summer of 2018, local branches of the Farmers Association invited farmers to talk about the crisis and to find support in sharing problems with each other. Farming can be a lonely occupation, so to provide an opportunity for the farmers to talk to and support each other, and feel connected to others was important. Although most farmers came through the crisis without suffering long-term consequences, a limited but unknown number of farmers experienced severe economic problems. One of the managers in the Farmers Association described their responses as follows in an interview:

> We [the Farmers Association] have the position that if we suspect that anyone are having a hard time, we should not be scared to invite ourselves for a cup of coffee. We are worried that those struggling are the ones we do not hear from.

Other institutions also took actions to reduce vulnerability in the agricultural sector. Banks offered economic advice to farmers and The Ministry of Agriculture removed customs on imported fodder. A government-operated relief scheme enabled farmers to seek economic compensation for their losses. Based on applications, farmers could receive a compensation covering up to 60 per cent of average sales income from grain. About 11,000 farmers in Eastern Norway received 150 million EUR in compensation from the government during autumn and winter 2018/19, with each farmer receiving on average about 13,000 EUR. This was important, since the compensation financed purchase of input for the following year's growing season. However, the compensation was not sufficient to offset a significant reduction in farm income.

While institutions helped to reduce vulnerability, these actions have to be contextualised within longer-term structural change in Norwegian agriculture. These structural changes have contributed to increased biophysical and economic vulnerability:

> Farmers recognize that structural changes in Norwegian agriculture during recent decades have influenced, and in certain areas increased, farmers' vulnerability. In order to use large tractors and machines more efficiently, fields have been enlarged by removing lines of trees that formerly divided them and putting small streams in pipelines.
>
> (Flemsæter et al. 2018, 2056)

Such structural changes have been supported by national agricultural policies since the 1960s (Almås 2002). These structural changes were part of the 'productivist' turn in agriculture across Europe in the aftermath of World War II (Spaargaren et al. 2012). Norway experienced annual productivity growth in the agricultural sector of four per cent after 1990, which is twice the productivity growth in industry (Ladstein and Skoglund 2005). Productivity growth and structural changes have provided the general population with relatively cheaper food. Specialisation in one or two products at the farm level has been part of such ongoing change. This has resulted in economies of scale and productivity growth, but it also increased vulnerability among farmers. Specialising in grain when the harvest fails due to weather conditions is one example of reduced resilience due to specialisation at the farm level. In the context of the 2018 drought, this specialisation meant farmers had to collaborate across the different divisions of labour within the farming community to reduce vulnerability.

Vulnerability reduction in the personal sphere

Perceptions play an important role in how one calculates future risks and assesses vulnerability (Slovic 2000), and it influences the long-term actions one takes to reduce risk and vulnerability in the political and practical spheres. Perceptions concerning the likelihood of a drought happening again provide different motivations for the actions deemed important to the informants. The summer of 2018 suggests that sustained periods of hot and dry summer weather are part of ongoing climate change across larger areas of northern Europe (Buraas et al. 2020). Future climate scenarios for Norway project temperature increases of about 4.5°C by the end of this century, together with increased rain, changes in precipitation patterns, shrinking glaciers, and rising sea levels (Hanssen-Bauer et al. 2017). However, as a signatory to the Paris Agreement on climate change, Norway is obliged to work to keep the global temperature increase well below 2°C. As a result, it is important for Norway to align climate change mitigation and adaptation policies with sustainable development of agriculture and food systems. Yet while climate change is widely discussed in politics and the media, responses to climate change have so far not been pronounced in the agricultural sector in Norway:

> ...even though many farmers see and reflect on the connections between climate change and agriculture, few take specific actions to adapt to or mitigate climate change. ... Farmers have taken a few actions voluntarily, mostly for economic reasons, and often connected to support schemes. There were some examples of ecologically motivated actions, but these seem to be rather rare.
>
> (Flemsæter et al. 2018, 2057)

Climate change mitigation through a reduction in greenhouse gas emissions needs to be integrated into Norwegian agricultural policies. There is a significant but largely unknown potential for carbon capture and storage in the topsoil by changing farming practices (Norwegian Environment Agency 2020). Through organic farming and other practices such as no till agriculture, the content of carbon in the topsoil will increase, relative to conventional farm practices such as ploughing. Based on current knowledge, optimising the use of nitrate-based fertiliser combined with planting more wheat and Canola in the autumn seems to be among the most promising ways forward concerning reduced emissions from grain production in Norway (Bardalen et al. 2018). Winter varieties of Canola did well during the 2018 drought and researchers in agronomy argue for more Canola production in Norway, which is very limited at present. Based on such advice, a voluntary agreement between government and the Farmers Association has been reached. The Farmers Association has established a vision for a fossil-free agricultural sector within 2030.

While this sounds promising, there is no consensus among farmers about the significance of climate change. Based on participatory observation, interviews, and newspaper articles, it became evident that there are two positions on how farmers make sense of the challenges of the summer of 2018, which influenced responses to the challenges. One position is based on the belief that 2018 was an exception and something that would not likely happen again, whereas the other position is that this was something farmers should be better prepared for in the future. The first position was summarised by an informant as follows:

> I believe 2018 was an extreme year, and that it will not happen again. The meteorologists explained that is was a very special weather situation with a blocking high pressure. I do not believe that the grain producers can be

> prepared for years that are this dry. If a year like this happens, it is important to reduce the costs by reducing fertilisers and pesticides.

This informant argued for the need to accept the losses and focus on limiting the economic damages by being critical towards which activities to prioritise. Actions such as using fertilisers and pesticides have high costs and are only prioritised when it is considered economically beneficial to do so. During the hot summer of 2018, farmers were advised to do nothing by the extension services. Irrigation was only prioritised for crops when it would be economically beneficial to do so.

The second position was formulated quite precisely by another grain farmer, who also had a managerial position within the Farmers Association:

> We need to be prepared that this can happen more frequently. We have to have a plan that is better than the one we had this year, it was a bit all over the place. We have to get coordinated earlier. The individual farmer has to build a buffer stock, for example with grass. The individual farmer has to have their own preparedness.

If actors within the agricultural sector believe that climate change poses risks to agriculture in Norway, and that adaptation and mitigation policies are related, it is more likely that political and practical actions will be taken to address the structural drivers of vulnerability, which includes addressing the sources of greenhouse gas emissions. The personal, political, and practical spheres are interlinked and interacting, and vulnerability reduction, or lack of such reduction, involves all three of them. Only the second position identified in this study acknowledges the need to prepare for increases in future climate risks such as drought. Although the precise number of farmers and other actors who see a need to be better prepared is unknown at the moment, this number is too small to be able to set the agenda for better preparedness and thus reduced vulnerability.

Conclusion: a vulnerable future?

The consequences of the drought of 2018 were reduced largely due to the institutional setup of the agricultural sector in Norway. Farmer helping farmer according to the cultural norm of 'dugnad' and with the support of farmer-owned cooperatives were highly important to overcome both economic and psychological consequences of 'the hottest summer ever.' A government-operated insurance scheme contributed significantly to ease the economic burden of the drought. A well-organised farming community able to cooperate with government agencies was important in this respect. However, the institutional setup was pushed to its limits and some of the limitations have been demonstrated.

In the aftermath of the drought, no systematic discussion has taken place in the farming sector concerning the root causes and drivers of vulnerability. 'The hottest summer ever' was followed by the summer of 2019 with good conditions for grain production across Eastern Norway. The months of May and June in 2019 were cool and wet, which resulted in grain yields ten per cent above average. The good yields in 2019 made it easier for farmers to forget the drought of 2018. There are no signs of plans to address extreme weather events in Norwegian agriculture. There have been talks, but they have not resulted in much more than support for developing better varieties and an advice to farmers to prepare themselves by storing fodder and seeds. The Farmers Association still supports the ongoing specialisation and structural change towards larger farms, and considerations about economies of scale triumph over concerns about increased vulnerability. Political initiatives to reduce meat consumption to mitigate CO_2 emissions and improve public health are met with resistance from large parts of the farming community. Measures to adapt to a changing climate in the practical sphere are, however, implemented when backed by government financial support. This includes support for ditches and reduction in ploughing in the autumn.

With the exception of some discussion about mitigation of climate change within the agricultural sector, no links were made to Norwegian fossil fuel policies, nor to addressing the larger drivers of climate change risks and vulnerability. The drought of 2018 is rapidly becoming an event of the past and it is easy and comfortable to forget about it. We suggest that changes such as those experienced in 2018 will affect agriculture in different ways, depending on the type of production, location, and potential for adaptation. Vulnerability will be experienced differently across scales, yet will be strongly influenced by social and economic changes within the agricultural sector. The future of Norwegian agriculture lies as much in collaboration and the quality of institutions as in the weather, and perceptions of the relationship between weather and climate change will influence whether the risks of climate change continue to increase in the decades ahead.

According to the UN Food and Agriculture Organization (FAO), agriculture and food systems are an important part of the climate solution, 'But they must transform through inclusive, multisectoral approaches that reduce emissions, draw down carbon, and boost climate resilience and adaptation' (FAO 2019). So far there are few signs of such a transformation in the agricultural sector in Norway. The lack of powerful measures to avoid increases in vulnerability in Norway seems to be part of a global trend of business as usual. This makes the extreme emission scenarios for climate change important to consider:

> Under the extreme emission scenario (RCP8.5), the frequency of droughts is projected to increase over the whole of Europe, with few exceptions: moderate increase over Switzerland, Hungary, Poland, Belarus, Lithuania, and central Scandinavia, and mixed tendencies over Iceland. The severity of droughts is projected to strongly increase over the southern third of Europe and over northernmost Scandinavia. Excluding central Iceland and southern Norway, the entire European continent will be affected by more frequent and severe droughts as the century passes.
>
> (Spinoni et al. 2018, 1732)

The potential for Norwegian farmers to increase production in a future with severe pressures on farmland in key agricultural areas in Europe is very limited. At the same time, Norway is becoming increasingly dependent on importing grain to feed its population and seeds for growing grain when drought hits Eastern Norway. This case study thus challenges the assumption that vulnerability is about 'those others' (e.g., poor people in economically developing countries). This is increasingly being revealed as a myth, and it is part of a larger vulnerability narrative that has allowed business as usual to continue for decades. The typical narrative communicated by national and international institutions working on vulnerability and disaster risk has largely approached climate change as a technical problem that requires tactics and measures to reduce the impacts of gradual or extreme events through adaptation policies and practices. Rather than transforming the underlying drivers of vulnerability, many strategies and programmes have focused on adapting to climate change impacts and risks through 'development as usual' (Eriksen et al. 2015). The measures taken within Norwegian agriculture described in this chapter are basically part of a business-as-usual scenario. This case study emphasises that responses to risk in a highly developed economy have been decoupled from the wider social and political context that generates vulnerability in the first place. Environmental risks have been treated as separate and distinct from social, economic, and political processes. Such a separation of environmental risks and social processes perpetuates fragmented approaches that have deepened rather than alleviated vulnerability to climate change.

In many cases, the concept of 'vulnerability' is still reduced to a superficial diagnosis that can be addressed through techno-managerial adaptations (Nightingale et al. 2020; O'Brien and Selboe 2015). In treating vulnerability as primarily a technical challenge, the structural and systemic factors are often ignored. In other words, the social, economic, political, and cultural relationships that perpetuate inequality, uneven development, exploitation of people and resources, concentration of power and wealth among fewer people and corporations, and the continued development and consumption of fossil fuels are generally not addressed. Reducing vulnerability calls not only for understanding what makes people vulnerable, but also addressing the multiple processes that are contributing to risk and vulnerability. In the absence of transformative change, it is likely that stories of 'the hottest summer ever,' will be told again and again.

References

Almås, Reidar. 2002. *Norges landbrukshistorie IV. Frå bondesamfunn til bioindustri*. Oslo: Samlaget.

Bankoff, Greg, Dorothea Hilhorst, and Georg Frerks. 2004. *Mapping Vulnerability. Disasters, Development and People*. London: Routledge.

Bardalen, Arne, Synnøve Rivedal, Anders Aune, Adam O'Toole, Finn Walland, Hanna Silvennoinen, Levina Sturite, et al. 2018. *Utslippsreduksjoner i norsk jordbruk. Kunnskapsstatus og tiltaksmuligheter*. Ås, Norway: NIBIO.

Bechmann, Marianne, and Johannes Deelstra. 2013. *Agriculture and Environment: Long Term Monitoring in Norway*. Oslo: Akademika.

Bougroug, Achraf, and Tor Kristian Ånestad. 2019. *God vekst i fastlands BNP i 2018*. Oslo: Statistisk Sentralbyrå. Accessed May 25, 2020. https://www.ssb.no/nasjonal-regnskap-og-konjunkturer/artikler-og-publikasjoner/god-vekst-i-fastlands-bnp-i-2018.

Buraas, Allan, Anja Rammig, and Christian S. Zang. 2020. 'Quantifying impacts of the 2018 drought on European ecosystems in comparison to 2003.' *Biogeosciences*, 17 no. 6: 1655–1672.

Eriksen, Siri, Tor Håkon Inderberg, Karen O'Brien, and Linda Sygna. 2015. 'Introduction: Development as usual is not enough.' In *Climate Change Adaptation and Development: Transforming Paradigms and Practices*, edited by Siri Eriksen, Tor Håkon Inderberg, Karen O'Brien and Linda Sygna, 1–18. London: Routledge.

Eriksen, Siri, and Elin Selboe. 2012. 'The social organisation of adaptation to climate variability and global change: The case of a mountain farming community in Norway.' *Applied Geography*, 33: 159–167.

Field, Christopher, B. Vicente Barros, Thomas F. Stocker, and Qin Dahe. 2012. *Managing the Risks of Extreme Events and Disasters to Advance Climate Change Adaptation. Special Report of the Intergovernmental Panel of Climate Change*. Cambridge: Cambridge University Press.

Fischer, Günther, Mahendra Shah, Harrij Velthuizen, and Freddy O. Nachtergaele. 2001. *Global Agro-Ecological Assessment for Agriculture in the 21st Century*. Laxenburg, Austria: International Institute for Applied Systems and Analysis.

Flemsæter, Frode, Hilde Bjørkhaug, and Jostein Brobakk. 2018. 'Farmers as climate citizens.' *Journal of Environmental Planning and Management*, 61 no. 12: 2050–2066.

Food and Agriculture Organization of the United Nations (FAO). 2019. *FAO at the UN Climate Change Conference COP 25*. Accessed June 30, 2020. http://www.fao.org/climate-change/international-fora/major-events/cop-25/en/.

Frerks, Georg, and Stephen Bender. 2004. 'Conclusion: vulnerability analysis as a means of strengthening policy formulation and policy practice.' In *Mapping Vulnerability. Disasters, Development and People*, edited by Greg Bankoff, Dorothea Hilhorst and Georg Frerks, 194–205. London: Routledge.

George, Alexander L., and Andrew Bennett. 2005. *Case Studies and Theory Development in the Social Sciences*. Cambridge, MA: MIT Press.

Hanssen-Bauer, I., E.J. Førland, I. Haddeland, H. Hisdal, S. Mayer, A. Nesje, and J.E.Ø. Nilsen. 2017. *Climate in Norway 2100 – a Knowledge Base for Climate Adaptation*. Oslo: NCCS Report No. 1/2017.

Hilhorst, Dorothea, and Greg Bankoff. 2004. 'Introduction: Mapping Vulnerability.' In *Mapping Vulnerability, Disasters, Development and People*, edited by Greg Bankoff, Dorothea Hilhorst and Georg Frerks, 1–10. London: Routledge.

Ladstein, Tove, and Tor Skoglund. 2005. *Utviklingen i norsk jordbruk 1950–2005*. Oslo: Statistisk sentralbyrå.

Leichenko, Robin, and Karen O´Brien. 2019. *Climate and Society. Transforming the Future*. Cambridge: Polity Press.

Nightingale, Andrea J., Siri Eriksen, Marcus Taylor, Timothy Forsyth, Mark Pelling, Andrew Newsham, Emily Boyd, E. et al. 2020. 'Beyond Technical Fixes: Climate solutions and the great derangement.' *Climate and Development*, 12 no. 4: 343–352.

Norwegian Environment Agency. 2020. 'Klimakur 2030.' Accessed May 20, 2020. https://www.miljodirektoratet.no/globalassets/publikasjoner/m1625/m1625_sammendrag.pdf.

O'Brien, Karen. 2018. 'Is the 1.5°C target possible? Exploring the three spheres of transformation.' *Current Opinion in Sustainability*, 31: 153–160.

O'Brien, Karen, Siri Eriksen, Linda Sygna, and Lars Otto Næss. 2006. 'Questioning complacency: Climate change impacts, vulnerability and adaptation in Norway.' *Ambio*, 35 no. 2: 50–56.

O'Brien, Karen, and Elin Selboe. 2015. 'Climate change as an adaptive challenge.' In *The Adaptive Challenge of Climate Change*, edited by Karen O'Brien and Elin Selboe, 1–23. Cambridge: Cambridge University Press.

O'Brien, Karen, Linda Sygna, and Jan Erik Haugen. 2004. 'Vulnerable or Resilient? A multi-scale assessment of climate impacts and vulnerability in Norway.' *Climatic Change*, 64: 193–225.

Piketty, Thomas. 2020. *Capital and Ideology*. Cambridge MA: The Belknap Press of Harvard University Press.

Ragin, Charles. 1994. *Constructing Social Research*. Thousand Oaks: Pine Forge Press.

Ribot, Jesse. 2014. 'Cause and response: Vulnerability and climate in the Anthropocene.' *Journal of Peasant Studies*, 41 no. 5: 667–705.

Sarkodie, Samuel Asumadu, and Vladimir Strezov. 2019. 'Economic, social and governance adaptation readiness for mitigation of climate change vulnerability: Evidence from 192 countries.' *Science of the Total Environment*, 656: 150 –164.

Gangstø Skaland, Reidun, Hervé Colleuille, Anne Solveig Håvelsrud Andersen, Jostein Mamen, Lars Grinde, Helga Therese Tilley Tajet, Elin Lundstad, et al. 2019. *Tørkesommeren 2018*. Oslo: Norwegian Meteorological Institute.

Slovic, Paul. 2000. *The Perception of Risk*. London: Routledge.

Smid, M., S. Russo, A.C. Costa, C. Granell, and E. Pebesma. 2019. 'Ranking European capitals by exposure to heatwaves and cold waves.' *Urban Climate*, 27: 388–402.

Spaargaren, Gert, Anne Loeber, and Peter Oosterveer. 2012. 'Food futures in the making.' In *Food Practices in Transition. Changing Food Consumption, Retail and Production in the Age of Reflexive Modernity*, edited by Gert Spaargaren, Peter Oosterveer and Anne Loeber, 1–23. London: Routledge.

Spinoni, Jonathan, Jürgen V. Vogt, Gustavo Naumann, Paulo Barbosa, and Alessandro Dosio. 2018. 'Will drought events become more frequent and severe in Europe?' *International Journal of Climatology*, 38 no. 4:1718–1736.

Statistics Denmark. 2019. *Crop production*. Accessed March 19, 2019. https://www.dst.dk/en/Statistik/emner/erhvervslivets-sektorer/landbrug-gartneri-og-skovbrug/vegetabilsk-produktion.

Statistics Norway. 2018. *Strukturen i jordbruket*. Accessed November 12, 2018. www.ssb.no/jord-skog-jakt-og-fiskeri/statistikker/stjord/aar/2018-12-06.

Statistics Norway. 2019. *Jord, skog, jakt og fiske: Jordbruk*. Accessed September 16, 2019. www.ssb.no/jord-skog-jakt-og-fiskeri?de=Jordbruk

Steffen, Will, Johan Rockström, Katherine Richardson, Timothy M. Lenton, Carl Folke, Diana Liverman, Colin P. Summerhayes, et al. 2018. 'Trajectories of the earth system in the anthropocene.' *PNAS*, 115 no. 33: 8252–8259.

Yu, Pei, Rongbin Xu, Michael J. Abramson, Shanshan Li, and Yuming Guo. 2020. 'Bushfires in Australia: A serious health emergency under climate change.' *The Lancet*, 4 no. 1: 7–8.

INDEX

Note: Page numbers in **bold** refer to tables and page numbers in *italics* refer to figures

Acts of God 34, 83, 192, 195
adaptation 2, 5, 15, 24–28, 161–165; climate change adaptation (CCA) 23–28, 143, 144, 150, 152, 161, 214, 215
Adaptive Social Protection (Shock Responsive Social Adaption) 143, 144, 161–163
Afghanistan 112, 116
African Union 148
aid programmes 19, 60–62, 114–120, 125–128, 145, 146, 152, 153; principles of 107–110, 126, 127, 129, 130; scenarios 112, 113; weaknesses 22, 114–116, 131–138, 179, 180, 200, 201; *see also* non-governmental organisations (NGOs)
Anthropocene epoch 196, 202, 206, 207
Artur, Luís 143–167
At Risk: Natural Hazards, People's Vulnerability and Disasters 3, 17
Australia 25, 223

Bangladesh 17, 53, 73, **74**, 75, 81, 174, 175
Bankoff, Greg 1–12, 15–32
beliefs, cultural and religious 76, 77, 82, 83, 192, 198
Bhopal disaster 33, 34, 38
Botswana 176, 177
Bradshaw, Sarah 51–67
Brazil 183, 184
Burton, Ian 192–205

Cameroon 184
Cannon, Terry 68–87
Chile 174
China 17, 81
civil defence 16, 41, 42
climate change 5, 72, 94, 95, 198–203, 208–212, 217; agriculture case study 222–235; climate change adaptation (CCA) 23–28, 143, 144, 150, 152, 161, 214, 215
Colombia 91, 92, 96, 98–104
community-based disaster risk reduction (CBDRR) 21, 22, 71
conflict situations 8, 9, 39–44, 91, 92, 99, 110–117; in Colombia 92, 98–101; in Mozambique 145; in Philippines 92, 97; in Sudan 125, 130–133, 135–138; women in 40, 52, 59–62
COVID-19 pandemic 19, 145, 201, 202
cultural perceptions of risk 69–71, 82, 83, 173
cyclones and typhoons 17, 73–75, 96–98, 145, 210; Idai 145; Kenneth 145; Nargis 81; Pablo 92, 96, 97, 102

Darfur *see Sudan*
DeLoughrey, Elizabeth 207, 211
development practices 16, 17–20, 25, 26, 148, 175, 200, 203; and women 55–59
Dickinson, Thea 192–205
disaster paradigms **195**, **196**

disaster risk creation (DRC) 7, 8, 185, 193, 197; responsibility 198–203
disaster risk reduction (DRR) 35, 185, 195–203, 208, 214, 215; omissions and inadequacies 36, 37, 43–45, 73–76, 81–83; trends and changes 15, 21, 22, 34, 71, 72; *see also* Hyogo Framework Sendai Framework
displaced persons *see* internally displaced persons (IDPs) refugees and migrants
droughts 153–155, 162, 209, 210, 213; European example 222–235

earthquakes 17, 20, 34, 36, 77, 79, 83, 110
El Salvador 22
Ethiopia 112, 113, 116
European Union (EU) 129, 136, 148

G20 148
Global Network of Civil Society for Disaster Reduction 186
Global Risk Report (WEF) 147
globalisation 23, 38, 146, 198, 202, 203

Haiti 36, 37, 60, 61, 112, 113, 116
health 35, 75–77, 80, 128, 129, 131, 183, 184
Hewitt, Kenneth 33–50, 93, 197
Hilhorst, Dorothea 1–12, 107–124
human influences 33, 34, 44, 45, 82, 83, 197–203
humanitarian aid *see* aid programmes
hurricanes 72, 73, 180, 181; Florence 180, 181; Katrina 36, 72, 73, 180, 181; Maria 180, 181, 186; Michael 180, 181; Mitch 22, 73
Hyogo Framework 8, 21, 28n3, 35, 108, 114, 200

India 75
indigenous peoples 98, 102, 184, 185
Indonesia 184, 185; Aceh 110
inequality 6, 60, 144, 145; gendered 53, 57
insurance 39, 81, 194, 233
Intergovernmental Panel on Climate Change (IPCC) 24, 25, 26, 200, 209
internally displaced persons (IDPs) 4, 127, 128; in conflict situations 133, 137, 145; in economic exploitation 184, 185; vulnerability of 181, 182
International Institute for Environment and Development 161
International Labour Organisation (ILO) 148, 149
Irish Aid 152, 161

Jaspars, Susanne 125–142

Keen, David 126, 138
Kelman, Ilan 206–221
Kenya 182
Kobe earthquake 20, 77

Linneker, Brian 51–67
local knowledge 176–179, 207, 208

Malawi 145
Mapping Vulnerability 2, 9
Maskrey, Andrew 21
Mika, Kasia 206–221
moral dimensions 43–45, 200, 201
mortality rates 17, 53, 57, *194*, *195*
mortality reduction 73–75
Mosse, David 22
Mozambique 143–147, 150–165, *153*
Myanmar/Burma 81, 112, 113, 116, 120

'natural' disasters 16, 68, 71–73, 75–78, 84n3, 94, 95, 103, 110, 152, 175, 176, 195–200
neoliberalism 4, 19–23, 24, 26, 27, 37, 77, 126, 137, 185
Nepal 36, 112, 113, 116, 177, 178
Nicaragua 22, 58
non-governmental organisations (NGOs) 20, 21, 61, 62, 71, 79, 115, 129, 130; *see also* aid programmes
Norway 222–235
nutrition/malnutrition 125, 128–138, 152

O'Brien, Karen 222–237
Oliver-Smith, Anthony 93
Overseas Development Institute 25
Overton, Lisa 51–67
Oxfam 61, 79

Pacific island nations 25
participatory approaches 21, 22, 79, 176
Pelling, Mark 25, 26
Peru 178, 179
Philippines 21, 91, 92, 96–98, 102–104
poverty: gendered 54–57, 155–157; rates 155; and vulnerability 36–39, 71, 152; in Western discourse 5, 15, 18, 20, 24; *see also* precarity
power relations 35, 79, 81, 93, 135, 164, 171–186; patriarchal 54, 60–63; state power 41, 91, 92, 95–104, 114–116, 171, 172, 175–178
precarity 4–6, 37–39, 180–182, 186

'pressure and release' (PAR) framework 17, 172, 173
Puerto Rico 180, 181, 186

reconstruction 42, 43, 58, 145, 200
Red Cross 72, 79, 81
refugees and migrants 4, 5, 108, 116–120, 127, 128, 129, 183
resilience approaches 1, 2, 9, 19–28, 62, 107–109, 114–117, 119, 120, 125–138; inadequacies 22, 23, 116, 117, 131–138, 147, 152
Ribot, Jesse 24, 25, 27

Sæther, Bjørnar 222–237
Sahel 161, 213
Schaffer, Bernard 126, 130
Sendai Framework 8, 21, 28n3, 35, 39, 114, 198, 199, 200, 203n1
Siddiqi, Ayesha 91–106
Sierra Leone 112, 113, 116
social protection (SP): in Mozambique **158**
social protection (SP) strategies 7, 143, 144, 148–150; Adaptive Social Protection (Shock Responsive Social Adaption) 143, 144, 161–163; in Mozambique 150–152, 157–165
social structures and contexts 2, 3, 16, 17, 33, 34, 131; and vulnerability concept 68–72, 82, 83, 94, 95
Somalia 173, 174, 183
South Sudan 112, 134, 179
Sri Lanka 53, 110
Sudan 117, 125–138
Syria 118

trade, international 3, 146, 147
tsunamis 36, 53, 110
Turkey 79, 118
typhoons *see* cyclones and typhoons

Uganda 99
United Nations: Framework Convention for Climate Change (UNFCCC) 24, 25, 26, 198–201, 209; High Commissioner for Refugees (UNHCR) 118, 120; International Children's Emergency Fund (UNICEF) 148, 175; International Decade for Natural Disaster Reduction (IDNDR) 3, 71; Office for Disaster Risk Reduction (UNDRR) 25, 72, 197, 198; Office for the Coordination of Humanitarian Affairs (OCHA) 127, 128
United States 180, 181; New Orleans 36, 72, 73, 181

Violence against Women and Girls (VAWG) 60, 61

vulnerability 1–10; components of 80–82; development of concept 15–19, 26–28, 36, 44, 45, 68, 69, 93–95; European example 222–235; and human behaviour 33–39, 76, 192–203; inadequacies of concept 147, 148, 215, 216; as overall approach 68–83, 207–218; and poverty 36–39, 71, 152; and resilience approach 107–109, 114–120; and social structures 68–72, 82, 83; of women 51–63, 118, 119

warfare *see* conflict situations
White, Gilbert 44, 68, 69, 192–193, 197, 200
Wisner, Ben 22, 39, 93, 148, 171–191
women 174, 175, 183, 184; in conflict situations 40, 52, 59–62; and development practices 55–59; and poverty 54–57, 155–157; and vulnerability 51–63, 118, 119
World Bank 22, 57, 58, 59, 72, 108, 148, 149, 161; reports 71, 79, 80, 131, 143, 144, 145
World Food Programme (WFP) 60, 125, 128, 131–135
World Health Organization (WHO) 76

Yemen 112

Zika virus 183, 184
Zimbabwe 112, 113